Contributions to Statistics

V. Fedorov / W. G. Müller / I. N. Vuchkov (Eds.) Model Oriented Data-Analysis, XII/248 Seiten, 1992

Jaromír Antoch (Ed.)

Computational Aspects of Model Choice

With 19 Figures

Physica-Verlag
A Springer-Verlag Company

Series Editor
Werner A. Müller
Peter Schuster

Editor
Prof. Dr. Jaromír Antoch
Department of Mathematical
Statistics and Probability
Charles University
Sokolovská 83
18600 Prague 8, CSFR

CIP-Titelaufnahme der Deutschen Bibliothek
Computational aspects of model choice / Jaromír Antoch (ed.)
- Heidelberg : Physica-Verl.; New York : Springer, 1992
(Contributions to statistics)
ISBN-13: 978-3-7908-0652-6 e-ISBN-13: 978-3-642-99766-2
DOI: 10.1007/978-3-642-99766-2
NE: Antoch, Jaromír [Hrsg.]

7100/7130-543210 - Printed on acid-free paper

PREFACE

Although no–one is, probably, too enthused about the idea, it is a fact that the development of most empirical sciences to a great extend depends of the development of data analysis methods and techniques, which, due to the necessity of applications of computers for that purpose, actually means that it practically depends on the advancements and orientation of computational statistics.

This volume contains complete texts of the lectures held during the Summer School on "Computational Aspects of Model Choice" organized jointly by Charles University, Prague, and International Association for Statistical Computing (IASC) on July 1 – 14, 1991, in Prague. Main aims of the Summer School were to review and analyse some of the recent developments concerning computational aspects of the model choice as well as their theoretical background. The topics covers the problems of the change point detection, robust estimation and its computational aspects, classification using binary trees, stochastic approximation and optimization including the discussion about available software, computational aspects of graphical model selection and multiple hypotheses testing. The bridge between these different approaches is formed by the survey paper about statistical applications of artificial intelligence.

While final preparations of the Summer School were getting underway, we were deeply saddened by the unexpected death of Tomáš Havránek, one of the initiators of this project, on May 17, 1991. Another very sad message came when preparing this volume. Bernd Streitberg died of a heart attack on September 28, 1991. Both left their contributions unfinished on their tables. While in our hearts their memory will endure, for those who did not meet them personally we have prepared at least short reminiscences complemented with concise bibliography of their most important results.

As the editor of this volume I would greatly acknowledge the fruitful cooperation of the authors and referees during the preparation of this book and Dr. P. Schuster for handling the volume in Physica Verlag, Heidelberg.

During the preparation of the Summer School we have benefited greatly from suggestions and encouragements of Prof. N. C. Lauro, president - elect of IASC. BMDP® Statistical Software, Inc., DATAVISION® AG and SAS® Institute, GmbH, deserves our thanks for the possibility to use their programs during the practical exercises. Many thanks go also to the Institute of Computer and Information Science of the Czechoslovak Academy of Sciences for their generous help. Computers were provided by the organizers, IBM Academic Initiative in Czechoslovakia and TIS a. s., Apple® Computer IMC Praha.

Finally, I must express my greatest attitude to P. Bitzan, Z. Brožková, P. Charamza, K. Chmel, K. Eben, J. Emmer, Z. Fabián, J. Hanousek, M. Malý, V. Šebesta and J. Á. Víšek for well done job during the preparation of the Summer School. Without their hard work we never could put our Summer School into the life.

Praha, May 1992 Jaromír ANTOCH

CONTENTS

TOMÁŠ HAVRÁNEK

JULY 14, 1947 — MAY 17, 1991

Tomáš Havránek was born in Prague in the family of well known bohemist academician B. Havránek. His carrier started in 1972 just after finishing Charles University and fulfilling military service. The first job has been that of statistician - consultant in the Institute of Microbiology of the Czechoslovak Academy of Sciences. Here he split interests into the two parts, the routine statistical analysis of biological data and his own scientific problems. And on this place he has found a lot of ideas for books, papers and lectures which followed soon.

His scientific interests were more then wide. The best witnesses are books, more than one hundred papers and technical reports he finished and numerous lectures for very heterogeneous auditoriums he delivered. Aside them he left behind a desk full of many others projects, unfortunately unfinished. Their subjects spread out from the mathematical and computational statistics and informatics up to the algebra, mathematical logic and the problems of the artificial intelligence. But he was not pure theoretician only. On the contrary, he has been proud that he knows the methods he is dealing with also from the side of the user.

His interest were not limited to statistics and mathematics only. Tomáš also loved teaching and during last twenty years supervised many theses. He likewise concentrated on history and his knowledge from this field has been enormous. In gymnasium he hesitated whether to study mathematics or history.

Tomáš took active part in many local conferences and seminars. And not only as the mere participant or speaker, but very often also as one of the organizers. He recognized soon that the preparation of the papers is only one part of a job. Therefore, he claimed and tried to assert his colleagues for, that it is necessary to introduce new ideas also to the users of the statistics, i. e., to those who do not open the

Prepared by Jaromír ANTOCH and his colleagues from Charles University, Prague, and Academy of Sciences, Prague .

Annals regularly but who perform everyday routine statistical analysis. This has been also one of the reasons why he initiated the origin of the Seminar on Applied Statistic within the Czechoslovak Union of Mathematicians and Physicists 16 years ago and why he prepared so many lectures for.

On the other hand, he did not concentrated his interests on small local conferences only. On the contrary, he took the active part also in the big ones, e. g., ISI and EMS meetings, BS Congress etc. But the key role played his involvement in the COMPSTAT meetings. He has been gradually the author of the contributed papers, member of SPC and finally invited speaker. Nevertheless, a lot of people remember him especially as the main organizer of COMPSTAT 84.

Success of COMPSTAT 84 had for Tomáš several consequences. On one side he had to change from the Institute of Microbiology to the Central Computing Center of the Czechoslovak Academy of Sciences (some people could not trust the success of the mathematicians in "purely biological" Institute). He continued here his work and has been soon appreciated by his new colleagues, most of them being the computer scientists. This appreciation can be illustrated especially on the fact that after November 1989 he has been elected to the post of the director of the Center and soon he succeeded to change it from the "mere" Central Computing Center to the Institute of Computer and Information Science of the Czechoslovak Academy of Sciences.

Aside that Czechoslovak computational statistics has been appreciated much more within IASC and Tomáš has been elected member of its Council for the period 1985 – 1989 and reelected again for the period 1991 – 1995. COMPSTAT 84 has been important also for the Czechoslovak statistical community, which could see what is new in the field and to present the results to the international statistical community.

But there has been another feature why we liked him – maybe more than for his scientific and organizational talent. This has been his humanity. Tomáš never did great difference between the ordinary student and university professor, between the inferior and superior. On the contrary, everybody was allowed to come at any time and to ask for a help (and this has not been withheld). And what we appreciated on him even more was his readiness to solve the problems and not to throw them over, as well as the confidence that we can rely on his words.

Tomáš Havránek passed away after short serious illness on May 17, 1991 and left behind wife and two sons. The news about his death was a regular thunderbolt and we could not believe to for a long time.

References

Books

[1] Havránek T. and Hájek P., *Mechanizing hypotheses formation - mathematical foundations of a general theory*, Universitexte, Springer-Verlag, Heidelberg, 1978.

[2] Havránek T., Albrecht V., Dvořák I., Jirků P. and Louvar B., *Matematika pro biologické a lékařské vědy (Mathematics for biological and medical sciences)*, Academia, Praha, 1981.

[3] Havránek T. and Hájek P., *Avtomaticeskoe obrazovanie gipotez*, Nauka, Moskva, 1984. (Russian translation of [1].)

[4] Havránek T., Hájek P. and Jiroušek R., *Processing uncertain information in expert systems*, RC Press, Boca Raton (to appear).

[5] Havránek T., *Statistika pro biologické a lékařské vědy (Statistics for biological and medical sciences)*, Academia, Praha (to appear).

Articles (excluding technical reports)

[6] Havránek T., *The statistical interpretation and modification of GUHA method*, Kybernetika **7** (1971), 13-21.

[7] Havránek T., *A generalization of the propositional calculus for purposes of the theory of logical nets with probabilistic elements*, Kybernetika **10** (1974), 13-43.

[8] Havránek T., *Statistical quantifiers in observational calculi*, Theory and Decision **6** (1975), 213-230.

[9] Havránek T., *Statistics and computability*, Kybernetika **12** (1976), 303-315.

[10] Havránek T., *Some aspects of automatic systems of statistical inference*, Transactions of 1974 European Meeting of Statisticians, vol. A, Academia, Praha, 1977, pp. 221-229.

[11] Havránek T. and Hájek P., *On generation of inductive hypotheses*, Int. J. of Man-Machine Studies **9** (1977), 415-438.

[12] Havránek T., *Towards a model theory of statistical theories*, Synthese **36** (1977), 441-458.

[13] Havránek T. and Hájek P., *The GUHA method, its aims and techniques*, Int. J. of Man-Machine Studies **10** (1978), 3-22.

[14] Havránek T. and Pokorný D., *GUHA-style processing of mixed data*, Int. J. of Man-Machine Studies **10** (1978), 47-57.

[15] Havránek T. and Pokorný D., *On some procedures for identifying sources of dependence in contingency tables*, COMPSTAT 78, Physica-Verlag, Wien, pp. 221-227.

[16] Havránek T., *On control of computer packages for data analysis*, The 2nd IFIP/IFAC Symp. on Software for Computer Control (Novák M., ed.), Pergamon Press, New York, 1979, pp. 300-307.

[17] Havránek T., *Approximate distribution of the maximum of 2x2 statistics derived from an RxC contingency table*, Proceedings of the Second Prague Symposium on Asymptotic Statistics, North Holland, Amsterdam, 1979, pp. 212-219.

[18] Havránek T., Radilová J. and Radil T., *A quantitative description of perception of the Necker cube*, Physiologia Bohemoslovaca **22** (1979), 427-428.

[19] Havránek T., *Alternative approach to missing information in the GUHA method*, Kybernetika **16** (1980), 145 – 155.

[20] Havránek T., *Some comments on GUHA procedures*, Explorative Datenanalyse, Medizinische Informatik und Statistik (Victor N., Lehmacher W. and van Eimeren W., eds.), vol. 26, Springer – Verlag, Heidelberg, 1980, pp. 156 – 177.

[21] Havránek T., *The present state of the GUHA software*, Int. J. Man – Machine Studies **15** (1981), 253 – 264.

[22] Hájek P. and Havránek T., *GUHA – 80. An application of artificial intelligence to data analysis*, Computer and Artificial Intelligence **1** (1982), 107 – 134.

[23] Havránek T., *On GUHA procedures for multidimensional contingency tables*, XI. Conf. Int. de Biometrie, Toulouse, 1982, pp. 41.

[24] Havránek T., *Some complexity considerations concerning hypotheses in multidimensional contingency tables*, Trans. 9th Prague Conf. on Inf. Theory, Statist. Dec. Functions and Random Processes, Academia, Praha, 1983, pp. 281 – 286.

[25] Havránek T. and Hájek P., *Logic, statistics and computers*, Logic in the 20th Century, Scientia, Milano, 1983, pp. 56 – 76.

[26] Havránek T. and Chytil M., *Mechanizing hypothesis formation – way for computerized exploratory data analysis*, Bull. Int. Statistist. Institute 44th ISI Meeting, Madrid, 1983, pp. 104 – 121.

[27] Havránek T., Lane P., Molenaar I., Nelder J. A., Tiit E. M., Verbeek A. and Victor N., *Standard packages versus tailor made software – panel discussion at COMPSTAT 84*, Statistical Software Newsletter **10** (1984), 56 – 57.

[28] Havránek T., *A note on the rank monotone dependence function*, Statistics **15** (1984), 369 – 372.

[29] Havránek T., *A procedure for model search in multidimensional contingency tables*, Biometrics **40** (1984), 95 – 100.

[30] Havránek T. and Lienert G. A., *Local and regional vs. global contingency testing*, Biom. J. **26** (1984), 483 – 494.

[31] Havránek T., Radilová J. and Radil T., *Sequential dependences of perceptual interpretations of a repetitively illuminated reversible figure*, Int. J. Psychophysiology **2** (1984), 45 – 50.

[32] Havránek T. and Jirků P., *A note on verbosity levels in cognitive problem solvers*, Computer and Artificial Intelligence **4** (1985), 15 – 20.

[33] Havránek T. and Edwards D., *A fast procedure for model search in multidimensional contingency tables*, Biometrika **72** (1985), 339 – 351.

[34] Havránek T. and Pokorný D., *On the GUHA approach to model search in connection to generalized linear models*, Generalized Linear Models, Lecture Notes in Statistic, vol. 32, Springer – Verlag, Heidelberg, 1986, pp. 82 – 92.

[35] Havránek T. and Lienert G. A., *Remission control of pre – post treatment comparisons by two – sample symmetry testing*, Methods of Information in Medicine **25** (1986), 116 – 122.

[36] Havránek T., *On general algorithm for model choice in multivariate analysis*, Proc. 7th Int. Summer School on Problems of Model Choice and Parameter Estimation in Regression Analysis, Sekt. Math. der Humboldt Univ., Berlin, 1986, pp. 88 – 98.

[37] Havránek T. and Lienert G. A., *Pre – post treatment evaluation by symmetry testing in square contingency table*, Biometrical Journal **28** (1986), 927 – 935.

[38] Havránek T. and Edwards D., *A fast model selection procedure for large families of models*, Journal of the Amer. Statist. Assoc. **82** (1987), 205 – 213.

[39] Havránek T. and Edwards D., *On variable selection and model choice in multivariate analysis*, DIANA II, MÚ ČSAV, Praha, 1987, pp. 161 – 174.

[40] Havránek T. and Hájek P., *A note on the independence assumption underlying subjective Bayesian updation in expert systems*, Artificial Intelligence and Information Control Systems of Robots, Elsevier, Amsterdam, 1987, pp. 41 – 47.

[41] Havránek T., *Model search in large model families*, Proceedings of the First World Congress of the Bernoulli Society, vol. 2, VNV Press, Utrecht, 1987, pp. 327 – 338.

[42] Havránek T., *Model search methods for contingency tables and intensional expert systems*, Trans. 10th Prague Conf. on Inf. Theory, Statist. Dec. Functions and Random Processes, Academia, Praha, 1988, pp. 375 – 384.

[43] Havránek T., *Comment on Streitberg's remarks on artificial intelligence in statistics*, Statistical Software Newsletter **14** (1988), 70 – 71.

[44] Havránek T., *On general algorithm for model choice in multivariate analysis*, Statistics **19** (1988), 465 – 476.

[45] Havránek T. and Strakoš Z., *On practical experience with parallel processing of linear models*, Bulletin of the International Statistical Institute **53** (1989), 105 – 117.

[46] Havránek T. and Soudský O., *Model choice in the context of simultaneous inference*, Statistical Data Analysis and Inference (Dodge Y., ed.), Elsevier, Amsterdam, 1989, pp. 165 – 176.

[47] Havránek T., *On Model Search Methods*, COMPSTAT 90 (Momirovič K. et Mildner V., eds.), Physica – Verlag, Heidelberg, 1990, pp. 101 – 108.

[48] Havránek T. and Jirků P., *Constructing an experimental expert system for exploratory data analysis*, Computational Statistics Quartely **5** (1990), 283 – 297, Physica – Verlag, Heidelberg.

[49] Havránek T., *Parallelization and symbolic computation techniques in model search*, SofStat 91, Gustav Fisher – Verlag, 1991, pp. 1 – 9.

TOMÁŠ HAVRÁNEK HAS BEEN AUTHOR OF MANY BOOKS, PAPERS AND TECHNICAL REPORTS. DUE TO THE LACK OF SPACE ONLY SOME OF THE MOST IMPORTANT AMONG THEM ARE LISTED ABOVE.

BERND STREITBERG

DECEMBER 18, 1946 — SEPTEMBER 28, 1991

Bernd Streitberg was born on December 18, 1946 in Bad Kissingen and belonged to an old Franconian family. He saw himself as a 'genuine' Franconian, but his favorite city was Berlin, where he had lived most of the time since he was a student. His early academic career was marked by the 1968 students' movement, which greatly influenced life in Berlin and which he enthusiastically supported. Since he was studying at the Institute of Sociology of the Free University and lived in the students' dormitory in Schlachtensee, he found himself at the center of these events. But his enthusiasm for different philosophies and brilliant argument of any kind precluded attempt to tie him to special political interests.

In mathematics, his other subject, he discovered his main passion and ability, if not to say his love for this kind of art. His topics ranged from applied statistics to differential equations and to abstract algebra. Influenced by empirical social studies, Bernd Streitberg soon came into contact with the newly developing field of computer sciences and information processing and thus he was stimulated to apply complex theoretical approaches to real data.

As an assistant fellow at the 'Correnspl atz', where the Institute of Applied Statistics of the faculty of Economics was located, he was a central figure on a team of young scientists. Their lives were filled with nightly discussions dedicated to mathematics, applied statistics and APL, a newly developed programming language of that period. Several well known scientists began their careers at this institute. Streitberg's extraordinary friendliness and helpfulness as well as his never–ending patience with people seeking advice, allowed him to play a central role in this community.

Prepared by Gisela ARNDT and her colleagues from Freie Universität Berlin.

A special event occurred in one of Streitberg's numerous lectures at that time, which were characterized by great clarity and a high mathematical level. He was impressed by a very gifted student, who periodically posed questions, which even Streitberg could not immediately answer. Afterwards, while drinking a beer together, this student told him he was a professor of mathematics. And so the Streitberg/Röhmel team was born, who provided so many new ideas about non-parametric methods and especially the shift-algorithm.

During the time at 'Correnspl atz', he also wrote many articles about time series analysis. This work finally led to a book about the topic, which was coauthored by Rainer Schlittgen. Today it has become the standard german textbook about time series analysis.

In 1983 Bernd Streitberg became Professor for Statistics and Computer Science at the Institute of Statistics and Econometrics of the Hamburg University. It was a very productive time. His publications, reports, courses and seminars covered a thematic field of unprecedented dimensions. His lectures were the highlight of many symposium. Their contents were well founded, they were excellently presented and frequently highly entertaining. With the breadth of his knowledge, his methodological competence, his curiosity about new and alternative ideas Streitberg became an ideal especially for many of the young colleagues. And accordingly, in 1990 he was elected president of the German Region of the International Biometric Society. He was the first scientist of his generation to reach this position. He used the strength of his personality to defuse generational conflicts. Moreover, he attempted to bring a breath of fresh air into the established structures of this ever-expanding society. The future will show the degree of his success.

He offered his advice on an impressive number of papers at many different institutes. His popularity as a mentor was not only due to his high degree of expertise, but also based on his ability to see almost every problem as an exciting one. This attitude was very motivating, however, during this he often remained in the background and introduced many ideas without taking credit. He was always very correct in quoting the original ideas of others.

Nor was his interest limited to mathematics and statistics. He likewise concentrated on philosophy (Paul Feyerabend), physics and music (programs for composing, articles on music theory, acoustic time series analysis). He was always able to bring mathematics into the realm of other subjects without intruding. Bernd Streitberg leaves behind a desk overflowing with unfinished projects. One publication – together with Samuel Kotz for the Annals – to which he attached great importance he had been able to complete. Others, unfortunately, will most likely remain unpublished.

Bernd Streitberg died of a heart attack on September 28, 1991 after a brilliant lecture at Prof. Kubicki's birthday symposium. He leaves behind his partner, Ms Kati Jauhiainen–Denzin, and their three year old daughter, Ionna, many bereaved friends and students. In our hearts his memory will endure.

REFERENCES

BOOKS

[1] Streitberg B., *Schätzung von Kovarianzstrukturen in linearen Zwei– und Mehrebenen– Modellen*, Dissertation, Freie Universität, Berlin, 1976.

[2] Streitberg B., *Über gruppentheoretische Methoden in der Varianzkomponenten Analyse*, Habilitation, Freie Universität, Berlin, 1981.

[3] Authors from several universities (Berlin, Bielefeld, Hannover and Kiel), *Statistik für Wirtschaftswissenschaftler*, Chapter on Wahrscheinlichkeitstheorie, Skriptum, Freie Universität Berlin, Berlin, 1979 (further editions followed).

[4] Buening H. et al., *Operationale Methoden in der Markt – und Sozialforschung*, Chapters on "Explorative Datenanalyse, Lineare Modelle, Multivariate Analyse and Loglineare Modelle", De Gruyter, Berlin, 1981.

[5] Streitberg B. and Schlittgen R., *Zeitreihenanalyse*, Oldenbourg Verlag, München, 1983 (1st edition). (5th edition in preparation)

[6] Streitberg B., *Mathematik für Wirtschaftswissenschaftler: Lineare Algebra*, Skriptum, Hamburg Universität, Hamburg, 1986.

ARTICLES (EXCLUDING TECHNICAL REPORTS)

[7] Streitberg B. and Birkenfeld W., *Analyzing Time Series in APL*, COMPSTAT 1976 (Gordesch J. and Naeve P., eds.), Physica Verlag, Wien, 1976, pp. 388 – 396.

[8] Streitberg B., *Multivariate Models of Dependent Spatial Data*, Exploratory and Explanatory Statistical Analysis of Spatial Data (Bartels and Ketellapper, eds.), Nijhoff, Leiden, 1979.

[9] Streitberg B., *Über eine Verallgemeinerung der Spektralanalyse*, Computational Statistics (Buening H. and Naeve P., eds.), De Gruyter, Berlin, 1981.

[10] Streitberg B., Röhmel J. and Herrmann W., *The COMSTAT Algorithm for Multimodal Factor Analysis – An Improvement of Tuckers Three Mode Factor Analysis Method*, Neuropsychobiology **10** (1983), 157 – 163.

[11] Streitberg B. and Röhmel J., *Exact Nonparametrics in APL*, ACM APL Quote Quad **14** (1984), 313 – 326.

[12] Streitberg B., *The Permutation Test Problem is NP – hard*, Statistical Software Newsletter **12** (1986), 18 – 19.

[13] Streitberg B. and Naeve P., *A Modestly Intelligent System for Estimation, Identification and Forecasting of Univariate Time Series*, Expert Systems in Statistics (Haux R., ed.), Fischer, Stuttgart, 1986.

[14] Streitberg B. et al., *COMSTAT Rule for Vigilance: Classification Based on Spontaneous EEG Activity*, Neuropsychobiology **17** (1987), 105 – 117.

[15] Streitberg B. and Röhmel J., *Exakte Verteilungen für Rang- und Randomisierungs- Tests im allgemeinen c – Stichprobenproblem*, EDV in Medizin und Biologie **18** (1987), 12 – 19.

[16] Streitberg B., *Expertensysteme in der Statistik – Experten oder Ignoranten?*, EDV in Medizin und Biologie **19** (1988). (Reprinted in ZUMA Software Conference Proceedings (1989))

[17] Streitberg B., *On the Nonexistence of Expert Systems – Critical Remarks on Artificial Intelligence in Statistics (with comments)*, Statistical Software Newsletter **14** (1988), 55 – 63.

[18] Streitberg B. and Herrmann W., *Vigilance Classification Based on the COMSTAT Rule*, Pharmacopsychiatry (1988).

[19] Streitberg B. and Röhmel J., *Exact Distributions for Nonparametric Tests against Partial Order Alternatives*, Computational Statistics Quarterly (1988), 23 – 41.

[20] Streitberg B. and Röhmel J., *Some Structural Aspects of Multiple Testing*, Multiple Hypotheses Testing (Bauer P., Hommel G. and Sonnemann E., eds.), Springer Verlag, New York, 1988, pp. 136 – 143.

[21] Streitberg B. and Balzer K., *The Sound of Mathematics*, Proceedings of the 14th International Computer Music Conference, Cologne, 1988, pp. 158 – 165.

[22] Streitberg B. and Röhmel J., *On Tests that are Uniformly More Powerful than the Nonrandomized Wilcoxon and Mann – Whitney U – test*, Biometrics **46** (1990), 481 – 484.

[23] Streitberg B., *Lancaster Interactions Revisited*, Annals of Statistics **18** (1990), 1878 – 1885.

Selected Technical Reports

[24] Streitberg B., *Prognose im Langenschen Verweilzeitmodell*, Diskussionsarbeit Nr. 7/79, Institut für Quantitative Ökonomik und Statistik, Freie Universität, Berlin, 1979.

[25] Streitberg B., *Vector Correlations of Time Series and the Box – Jenkins Approach to ARMA Identification*, Diskussionsarbeit 4/82, Institut für Quantitative Ökonomik und Statistik, Freie Universität, Berlin, 1982.

[26] Streitberg B. and Röhmel J., *Zur Konstruktion globaler Tests*, Preprint No 148, Fachbereich Mathematik, Freie Universität, Berlin.

[27] Streitberg B., *A Graphical Calculus of Multidimensional Arrays with an Application to Three Mode Factor Analysis*, Technical Report 135, Fakultät für Wirtschaftswissenschaften, Universität Bielefeld, 1985.

[28] Streitberg B., *The Metric Approach to the Construction of Experimental Designs. Part I: Invariant Metrics on Factorial Design Spaces*, Technical Report 154, Fakultät für Wirtschaftswissenschaften, Universität Bielefeld, 1985.

Bernd Streitberg has been author of many books, papers and technical reports. Due to the lack of space only some of the most important among them are listed above.

CHANGE POINT PROBLEM

JAROMÍR ANTOCH, MARIE HUŠKOVÁ

Charles University, Prague

ABSTRACT. The main objective of this paper is to survey procedures that one connected with the change point problem and with testing the constancy of regression relationship over a time. We will focus on the case of independent observations. Detailed algorithms are presented for most of the discussed methods.

1. INTRODUCTION

Let $X_1, \dots, X_n$ be independent random variables observed at ordered time points $t_1 < \dots < t_n$. Denote by F_i distribution function (d.f.) of X_i and assume that all the F_i belong to a common family $\mathcal{F}$. The problem of testing

$$H_0 : \quad F_1 = \dots = F_n \tag{1.1}$$

against

$$H_1 : \quad \text{there exists } 1 \le m < n \text{ such that}$$

$$F_1 = \dots = F_m \ne F_{m+1} = \dots = F_n,$$

together with the problem of estimation of m, is known as *change point problem*.

Let $X_1, \dots, X_n$ follow the linear model

$$X_i = \underset{\sim}{c}_i' \underset{\sim}{\theta}_i + e_i, \quad i = 1, \dots, n, \tag{1.2}$$

where $\underset{\sim}{c}_i = (c_{i1}, \dots, c_{ip})'$ are known regression vectors, $\underset{\sim}{\theta}_i = (\theta_{i1}, \dots, \theta_{ip})'$ are unknown (regression) vector parameters and $e_1, \dots, e_n$ are

1990 *Mathematics Subject Classification*. 62F35, 62G10, 62F03.

***Key words and phrases*. MOSUM, CUSUM, testing constancy of the regression relationships over a time, change point problem.**

independent identically distributed (i.i.d.) random errors with common distribution function (c.d.f.) F fulfilling certain regularity conditions and unknown otherwise. Then *testing the constancy of the regression relationships over a time* is formulated as follows, i.e.,

$$H_0^* : \quad \underset{\sim}{\theta}_1 = \cdots = \underset{\sim}{\theta}_n \tag{1.3}$$

against

$$H_1^* : \quad \text{there exists } 1 \le m < n \text{ such that}$$

$$\underset{\sim}{\theta}_1 = \cdots = \underset{\sim}{\theta}_m \neq \underset{\sim}{\theta}_{m+1} = \cdots = \underset{\sim}{\theta}_n.$$

The alternative hypothesis H_1 means that the first m observations have the same d.f. F_1 and starting from the $(m+1)$-st observation they have d.f. F_n. Analogously, the alternative hypothesis H_1^* says that the observations X_i for $i = 1, ..., m$, m unknown, follow the regression model with the vector parameter $\underset{\sim}{\theta}_1$ and then they switch to the regression model with a different parameter $\underset{\sim}{\theta}_n$. This means that there exists a time point $\tau \in (t_m, t_m + 1]$, usually called *change point*, in which we can observe the change of the regression relationship. A related problem of interest is to estimate τ.

There exist a lot of modifications of the above mentioned problems. For example, one can admit more then one change. Another possibility is that the change is continuous in the sense that it starts at a time point τ_1 and proceeds till some time point τ_2 ($> \tau_1$). The observations need not be independent.

These problems arise in many applications. Well known is the testing for stability of some critical product characteristics within statistical proess control. In meteorology, one often meets the problem to signal possible changes in climate or in weather. One can be interested in identification of economic structural changes. There is also a number of applications in biology and medicine, e.g., detection of a change in behavior or state of a subject etc.

There were developed quite a number of test procedures for the above formulated problems. According to the approach they can be classified into several groups, which will be concisely characterized in sequel.

1.A. Sequential and non-sequential approaches.

The sequential approach assumes that one has an infinite sequence $\{X_1, X_2, \dots\}$ of random variables gathered over another sequence $\{t_1, t_2, \dots\}$ of ordered points such that for some integer m, possibly equal to $+\infty$, $F_1 = ... = F_m \neq F_{m+1} = ...$ In this context one wishes to choose a ***stopping number*** N such that:

1. If m is finite, then the excess of N over m should be small.
2. If $m = \infty$, then the probability of a false alarm, i.e., $N < \infty$, should be small. This approach is worked out by Shiryayev (1963, 1978).

More attention was paid to so called ***quasi – sequential*** approach, which can be characterized as follows:

1. An upper bound for the sample size N_0 is prechosen.
2. After the i – th observation, $i \leq N_0$, one can either reject H_0 (or H_0^*) and stop or realize a new observation.
3. The final decision is done no later then after the N_0 – th observation. This means that the sample size is random and H_0 will be rejected (or not rejected) no later then after the N_0 – th observation.

If the sample size is not random, one speaks about a non – sequential approach.

1.B. Bayesian and non – Bayesian approaches.

Bayesian approach assumes that the change point τ is a random variable with the distribution

$$P(\tau \in (t_m, t_{m+1}]) = p_m, \quad m = 1, ..., n-1 \quad \text{and} \quad p_n = P(\tau > t_n).$$

The prior distribution expresses our knowledge (information) about τ. For further information see Ferreira (1975), Chin Choy and Broemling (1980), Broemling and Tsurumi (1986), Zacks (1983) or Krishnaiah and Miao (1988). The non – bayesian approach works with a constant unknown change point.

1.C. Classical and robust approaches.

When we assume that the distribution functions F_i are known up to parameters and the test statistic is constructed via the likelihood ratio method, we usually speak about ***classical approach***. Often the normal distribution is assumed and then the test statistic can be expressed via the least squares estimators.

It is known that the quality of the test procedures based on the least squares estimator are highly sensitive to deviations from normality, to heavy – tailed distribution, and to influential observations. For this reason so called ***robust estimators*** were developed which behave reasonably well and are rather insensitive to deviations from the assumed model. Typical robust procedures are based on M–, R– and L– statistics, or on Kolmogorov – Smirnov type statistics. The same arguments lead to the need of robust procedures for our testing problem. These can be based on either of the mentioned type of robust statistics. While the *non – recursive procedures* assume that we have at our disposal all observations at the beginning, the *recursive ones* belong to the quasi – sequential procedures.

A host of papers and some books were written about this problem, the well known survey papers of Csörgö and Horváth (1987, 1988), Deshayes and Picard (1986), Hackl and Westlund (1989), Hušková and Sen (1989), Krishnaiah and Miao (1988), Shaban (1980), Wolfe and Schechtman (1984) or Zacks (1991), e. g. The most popular paper on this problem was published by Brown, Durbin and Evans (1975). There exist also books and proceedings devoted to this problem, see, e.g., Basseville and Benveniste (1986), Broemling and Tsurumi (1987), Hackl (1980, 1990) or Schulze (1986) among others.

In our contribution we will concentrate on the robust procedures for testing the constancy of the regression relationships over time. One should notice that if in (1.2) $p = 1$ and $c_{11} = \cdots = c_{n1=1}$. The problem reduce to the change – point problem. In Sections 2 and 3 classical non – recursive and recursive procedures are briefly introduced. Section 4 gives basic information on M– estimators. Robust procedures are described in Section 5 (non – recursive) and in Section 6 (recursive).

1.D. Algorithms.

It is almost impossible to find programs of the mentioned methods in statistical packages at the mere touch of the button. On the other hand, it is evident that the implementation is relatively simple. Of course, the time needed for programming will be shorter in systems which offer basic building tools for the construction of macros like, e. g., *S+* or *ISP*. These tools should cover not only basic build - in functions, computation of quantiles and classical least squares estimator, but one would appreciate also minimization procedures, simple looping facilities, matrix operations etc. Huge statistical packages like *BMDP*, *SAS* or *SPSS* are useful for that purpose. While the methods can be relatively easily implemented by the *SAS* due to its powerful programming facilities and matrix language, much more time and tricks will be necessary to prepare analogous input file for *BMDP* or include file for *SPSS*.

To illustrate our ideas in more details, three simple algorithms covering non-recursive, CUSUM and MOSUM methods based on the classical least squares method are presented in Sections 2 and 3. The robustification of them, described in Sections 4 and 5, is quite obvious. To make the algorithms easily understandable, we use the pseudo Pascal notation and refer to the formulas only through the corresponding reference numbers.

2. CLASSICAL NON-RECURSIVE PROCEDURES

2.A. Classical procedures for shift in location.

In this section we present a survey of classical tests for the simple situation of a shift in location in order to give a flavor what type of test procedures were developed for the considered problem.

Let $X_1, \ldots, X_n$ be independent random variables fulfilling (1.2) with $c_{i1} = 1, i = 1, \ldots, n,\ p = 1$ and let the errors e_i be $N(0,1)$ distributed. The likelihood ratio method leads to the test with critical region given by

$$\max_{1 \le k \le n} \left\{ \frac{n}{k(n-k)} \left| \sum_{i=1}^{k} (X_i - \overline{X}_n) \right|^2 \right\} =$$

$$= \max_{1 \le k \le n} \left\{ \frac{nk}{n-k} (\overline{X}_k - \overline{X}_n)^2 \right\} > c_\alpha,$$

where c_α is determined such that the level is approximately α, $\overline{X}_k = k^{-1} \sum_{i=1}^{k} X_i$ and $\overline{X}^*_{n-k} = (n-k)^{-1} \sum_{i=k+1}^{n} X_i$.

Notice that $\frac{n}{k(n-k)} \left| \sum_{i=1}^{k} (X_i - \overline{X}_n) \right|^2$ follows, under H_0, χ^2 distribution with 1 degree of freedom,

$$\mathsf{E}_{H_0} \frac{k(n-k)}{n} (\overline{X}_k - \overline{X}^*_{n-k})^2 = 1$$

and

$$\mathsf{E}_{H_1} \frac{k(n-k)}{n} (\overline{X}_k - \overline{X}^*_{n-k})^2 = \begin{cases} 1 + (\theta_1 - \theta_n)^2 \frac{k(n-m)^2}{n(n-k)}, & k \le m, \\ 1 + (\theta_1 - \theta_n)^2 \frac{m^2(n-k)}{nk}, & k > m. \end{cases}$$

Clearly, the maximum of the expectation under H_1 is reached for $k = m$.

The exact critical value c_α can hardly be found. The Bonferroni inequality leads to $c_\alpha = \Phi^{-1}\left(1 - \frac{\alpha}{2(n-1)}\right)^2$. The resulting test is conservative. James et al. (1987) proposed critical region

$$\bigcup_{k=[nt_0]}^{[nt_1]} \left\{ \sqrt{\frac{n}{k(n-k)}} \left| \sum_{i=1}^{k} (X_i - \overline{X}_n) \right| > c^*_\alpha \right\},$$

where $0 < t_0 < t_1 < 0$, t_0 and t_1 are usually chosen close to zero and one, respectively, and c_α^* fulfills

$$\frac{c_\alpha^* \exp\{-c_\alpha^{*2}/2\}}{\Gamma(1/2)} \left\{ \frac{2}{c_\alpha^{*2}} + \frac{1}{2}\left(1 - \frac{1}{c_\alpha^{*2}}\right) \log \frac{t_1(1-t_0)}{t_0(1-t_1)} \right\} = \alpha.$$

2.B. Classical procedures for linear model.

Let $X_1, \dots X_n$ follow the linear model (1.2) with independent normally distributed random errors, i.e., X_i follows normal distribution $\mathcal{N}(\underset{\sim}{c}_i' \underset{\sim}{\theta}_i, \sigma^2)$, $i = 1, \dots, n$. The likelihood ratio method leads (for known σ^2) to the test statistic

$$\max_{p<k<n-p} V_{n,k}, \tag{2.1}$$

where

$$\begin{aligned} V_{n,k} &= (\hat{\underset{\sim}{\theta}}_k - \hat{\underset{\sim}{\theta}}^*_{n-k})'(\underset{\sim}{C}_k^{-1} + \underset{\sim}{C}^{*-1}_{n-k})^{-1}(\hat{\underset{\sim}{\theta}}_k - \hat{\underset{\sim}{\theta}}^*_{n-k})\sigma^{-2} \\ &= (\hat{\underset{\sim}{\theta}}_k - \hat{\underset{\sim}{\theta}}_n)'(\underset{\sim}{C}_k^{-1} - \underset{\sim}{C}_n^{-1})^{-1}(\hat{\underset{\sim}{\theta}}_k - \hat{\underset{\sim}{\theta}}_n)\sigma^{-2}. \end{aligned} \tag{2.2}$$

$\hat{\underset{\sim}{\theta}}_k$ and $\hat{\underset{\sim}{\theta}}^*_{n-k}$ in (2.2) are the least squares estimators of $\underset{\sim}{\theta}$ based on $X_1, \dots, X_k$ and $X_{k+1}, \dots, X_n$, respectively, i.e., for $p < k < n - p$

$$\hat{\underset{\sim}{\theta}}_k = \underset{\sim}{C}_k^{-1} \sum_{i=1}^{k} \underset{\sim}{c}_i X_i \quad \text{and} \quad \hat{\underset{\sim}{\theta}}^*_{n-k} = \underset{\sim}{C}^{*-1}_{n-k} \sum_{i=k+1}^{n} \underset{\sim}{c}_i X_i, \tag{2.3}$$

where

$$\underset{\sim}{C}_k = \sum_{i=1}^{k} \underset{\sim}{c}_i \underset{\sim}{c}_i' \quad \text{and} \quad \underset{\sim}{C}^*_{n-k} = \sum_{i=k+1}^{n} \underset{\sim}{c}_i \underset{\sim}{c}_i'. \tag{2.4}$$

Under H_0^* the distribution of statistic $V_{n,k}$ is χ^2 with p degrees of freedom. It is a great problem to find out the critical values. Usually, one takes some appropriate approximation. The simplest possibility is to make use of the Bonferroni inequality, which leads to the critical value

$$\max_{p<k<n-p} V_{n,k} > \chi^2_{\alpha_0}(p), \tag{2.5}$$

where $\chi^2_{\alpha_0}(p)$ is the $(1-\alpha_0)$-quantile of χ^2 distribution with p degrees of freedom and $\alpha_0 = \dfrac{\alpha}{2(n-p)}$. The level of this test is smaller than α.

Making use of the results of James et al. (1987) one receives the critical region (with asymptotic level α)

$$\max_{[nt_0]\le k\le[nt_1]} V_{n,k} > d_{\alpha,p}(t_0,t_1), \tag{2.6}$$

where $0 < t_0 < t_1 < 1$ and $d_{\alpha,p}(t_0,t_1)$ is the solution of the equation

$$\frac{b^p \exp\{\frac{-b^2}{2}\}}{2^{\frac{(p-2)}{2}}\Gamma(\frac{p}{2})}\left\{\frac{2}{b^2}+\frac{1}{2}\left(1-\frac{p}{b^2}\right)\ln\frac{(1-t_0)t_1}{(1-t_1)t_0}\right\} = \alpha. \tag{2.7}$$

If σ^2 is unknown, one replaces it by its estimator, e.g., by

$$\hat{\sigma}^2_n = \frac{1}{n-p}\sum_{i=1}^{n}(\underset{\sim}{X}_i - \underset{\sim}{c}_i'\hat{\underset{\sim}{\theta}}_n)^2, \tag{2.8}$$

and uses the same critical value as above.

The following formulae, i.e.,

$$\underset{\sim}{C}_k^{-1} = \underset{\sim}{C}_{k-1}^{-1} - \frac{\underset{\sim}{C}_{k-1}^{-1}\underset{\sim}{c}_k\underset{\sim}{c}_k'\underset{\sim}{C}_{k-1}^{-1}}{1+\underset{\sim}{c}_k'\underset{\sim}{C}_{k-1}^{-1}\underset{\sim}{c}_k}$$

and

$$\hat{\underset{\sim}{\theta}}_k = \hat{\underset{\sim}{\theta}}_{k-1} + \underset{\sim}{C}_k^{-1}\underset{\sim}{c}_k(X_k - \underset{\sim}{c}_k'\hat{\underset{\sim}{\theta}}_{k-1})$$

are useful for computations.

We illustrate the described method by means of a simple algorithm, in which $\underset{\sim}{T}$ is vector used for storing values of the statistics $V_{n,k}$ that are obtained in the individual steps of the procedure.

Algorithm 1. Classical non – recursive procedure.

{ SET UP }

$\underset{\sim}{T} \leftarrow \{0,\ldots,0\}$

Calculate $\hat{\underset{\sim}{\theta}}_n$, $\underset{\sim}{C}_n^{-1}$ and $\hat{\sigma}^2_n$ according to (2.3), (2.4) and (2.8).

{ PROCEDURE }

FOR $k := p+1$ TO $n-p-1$ DO

Calculate $\hat{\underset{\sim}{\theta}}_k$, $\underset{\sim}{C}_k^{-1}$ and $V_{n,k}$ according to (2.3), (2.4) and (2.2).

$\underset{\sim}{T}[k] := V_{n,k}$

$TestStat \leftarrow \max_{p<k<n-p} \underset{\sim}{T}[k]$

$k_{TestStat} \leftarrow \arg\max_{p<k<n-p} \underset{\sim}{T}[k]$

RETURN $TestStat$, $k_{TestStat}$, $\underset{\sim}{T}$

{ TEST }

IF $TestStat > \chi^2_{\alpha_0}(p)$

THEN We reject H_0^* on significance level α_0.

ELSE We cannot reject H_0^* on the significance level α_0.

PLOT $\{(i, \underset{\sim}{T}[i]),\ i = p+1, \ldots, n-p-1\}$

Remarks:

1. It is evident that the use of the critical value $d_{\alpha,p}(t_0, t_1)$ instead of $\chi^2_{\alpha_0}(p)$ in the TEST – part of the Algorithm 1. can affect only the statistical interpretation and not the behavior of the algorithm. Of course, in such a case we must change the FOR loop as follows, FOR $k := [nt_0]$ TO $[nt_1]$ DO … and the IF must be changed adequately.
2. If we use any of the robust estimators described in Section 4. instead of the classical least squares one, the only difference in Algorithm 1. will concern the lines starting with *Calculate* and the choice of the corresponding critical value. By other words, we should use either of the statistics (5.2) – (5.4) instead of (2.2), (5.6) instead of (2.8) and (5.9) – (5.11) instead of (2.5) or (2.6).
3. It is strongly recommended not only to take a look on the values of test statistics, but also to plot them against index (see the PLOT part of the algorithm).

3. CLASSICAL CUSUM AND MOSUM PROCEDURES

3.A. Classical procedures for shift in location.

Consider the same situation as in the Section 2.A. Recursive tests of the quasi – sequential type are based on the recursive residuals

$$W_k = \frac{X_k - \overline{X}_{k-1}}{\sqrt{\mathrm{var}_{H_0}(X_k - \overline{X}_{k-1})}} = \sqrt{\frac{k-1}{k}}(X_k - \overline{X}_{k-1}), \qquad k = 2, \ldots, n. \tag{3.1}$$

The random variables $W_2, \dots, W_k$ are independent, W_k follows normal distribution with unit variance,

$$\mathsf{E}_{H_0} W_k = 0, \qquad k = 2, \dots, n,$$

and

$$\mathsf{E}_{H_1} W_k = \begin{cases} 0, & k \le m, \\ -\frac{m}{\sqrt{k(k-1)}}(\theta_1 - \theta_n), & k > m. \end{cases}$$

The CUSUM (cumulative sum) procedures are based on the statistics $\sum_{k=2}^{t} W_k, t = 2, \dots, n$, and the critical regions are of the form

$$\bigcup_{t=2}^{n} \left\{ \left| \sum_{k=2}^{t} W_k \right| > w(\alpha; t, n) \right\}, \tag{3.2}$$

where $w(\alpha; t, n)$ are chosen such that the level is approximately α. The Bonferroni inequality leads to

$$w_1(\alpha; t, n) = \sqrt{t-1}\, \Phi^{-1}\left(1 - \frac{\alpha}{2(n-1)}\right), \qquad 1 < t \le n. \tag{3.3}$$

The resulting test is conservative. Brown et al. (1975) developed approximation

$$w_2(\alpha; t, n) = h_\alpha \frac{2t + n - 3}{\sqrt{n-1}}, \qquad 1 < t \le n, \tag{3.4}$$

where h_α fulfils

$$1 - \Phi(3h_\alpha) + \Phi(h_\alpha) e^{-4h_\alpha^2} = \frac{\alpha}{2}. \tag{3.5}$$

Notice that $h_{0,01} = 1,143$; $h_{0,05} = 0,948$ and $h_{0,1} = 0,85$.

The MOSUM (moving sum) procedures are based on statistics

$$\sum_{k=t-G+1}^{t} W_k, \quad t = G, \dots, n,$$

and the critical regions are of the form

$$\bigcup_{t=G+2}^{n} \left\{ \frac{1}{\sqrt{G}} \left| \sum_{k=t-G+1}^{t} W_k \right| > m(\alpha; G, k, n) \right\}, \tag{3.6}$$

where G is a suitably chosen integer. This means that G/n should be small, e.g., $G \in [0,1n; 0,2n]$, while $m(\alpha; t, n)$ should be chosen such that the level is approximately α. The Bonferroni inequality gives

$$m_1(\alpha; G, k, n) = \sqrt{G}\, \Phi^{-1}\left(1 - \frac{\alpha}{2(n-G-1)}\right); \tag{3.7}$$

a modified result of Deheuvels and Revész (1987) gives another critical region which will be discussed in Section 6.

3.B. Classical procedures for linear models.

Consider the same assumptions as in Section 2.B. $CUSUM$ procedures are based on the recursive residuals

$$\begin{aligned} W_k &= \frac{X_k - \underset{\sim}{c}_k' \hat{\underset{\sim}{\theta}}_{k-1}}{\sqrt{\mathsf{var}\,(X_k - \underset{\sim}{c}_k' \hat{\underset{\sim}{\theta}}_{k-1})}} = \\ &= \frac{X_k - \underset{\sim}{c}_k' \underset{\sim}{C}_{k-1}^{-1}(\underset{\sim}{c}_1, \dots, \underset{\sim}{c}_{k-1}) \underset{\sim}{X}_{k-1}}{\sigma\sqrt{1 + \underset{\sim}{c}_k' \underset{\sim}{C}_{k-1}^{-1} \underset{\sim}{c}_k}}, \quad k = p+1, \dots, n, \end{aligned} \tag{3.8}$$

where $\hat{\underset{\sim}{\theta}}_k$ is defined by (2.3) and $\underset{\sim}{X}_{k-1} = (X_1, \dots, X_{k-1})'$. Under H_o the random variables $W_{p+1}, \dots, W_n$ are i.i.d. according to $\mathcal{N}(0,1)$.

The $CUSUM$ (cumulative sums) test procedures are based on the statistics

$$M_t = \frac{\sigma}{\hat{\sigma}_t} \sum_{k=p+1}^{t} W_k = \frac{\hat{\sigma}_{t-1}}{\hat{\sigma}_t} M_{t-1} + \frac{\sigma}{\hat{\sigma}_t} W_t, \qquad t = p+1 \dots n, \tag{3.9}$$

where $M_p = 0$ and $\hat{\sigma}_t^2$ is an estimator of σ^2 based on $X_1, \dots, X_t$, e.g.,

$$\begin{aligned} \hat{\sigma}_t^2 &= \frac{1}{t-p} \sum_{i=p+1}^{t} \frac{\left(X_i - \underset{\sim}{c}_i' \hat{\underset{\sim}{\theta}}_{i-1}\right)^2}{1 + \underset{\sim}{c}_i' \underset{\sim}{C}_{i-1}^{-1} \underset{\sim}{c}_i} = \\ &= \frac{t-p-1}{t-p} \hat{\sigma}_{t-1}^2 + \frac{1}{t-p} \frac{\left(X_t - \underset{\sim}{c}_t' \hat{\underset{\sim}{\theta}}_{t-1}\right)^2}{1 + \underset{\sim}{c}_t' \underset{\sim}{C}_{t-1}^{-1} \underset{\sim}{c}_t}. \end{aligned} \tag{3.10}$$

The corresponding critical region is of the form

$$\bigcup_{t=p+1}^{n} \left\{ \frac{\sigma}{\hat{\sigma}_t} \left| \sum_{k=p+1}^{t} W_k \right| > w(\alpha; t, n) \right\}, \tag{3.11}$$

where $w(\alpha; t, n)$ is chosen in such a way that the level is (approximately) α. They are not uniquely determined. The main reason is that different (asymptotic) arguments lead to the different sets of critical values.

The Bonferroni inequality gives

$$w_1(\alpha; t, n) = \sqrt{t-p}\, t_{1-\alpha_1}(t-p), \tag{3.12}$$

where $t_{1-\alpha_1}(t-p)$ is the $(1-\alpha_1)$-quantile of the t-distribution with $(t-p)$ degrees of freedom and $\alpha_1 = \dfrac{\alpha}{2(n-p)}$. Notice that the corresponding level of the test is smaller than α.

Brown et al. (1975) recommended to use

$$w_2(\alpha; t, n) = h_\alpha \frac{2t+n-3p}{\sqrt{n-p}} \tag{3.13}$$

with h_α fulfilling (3.5). For other approaches see, e. g., Hackl (1980) or Hušková (1991 a).

Let us illustrate the CUSUM method in the form of an algorithm in which $\underset{\sim}{T}$ is the vector (of length n) used for storing values of the test statistics corresponding to the respective steps of the procedure. Notice that $\hat{\underset{\sim}{\theta}}$ is the auxiliary variable used in each step to keep the previous value of the estimator of $\underset{\sim}{\theta}$ needed for calculation of the statistic.

Algorithm 2. Classical CUSUM procedure.

{ SET UP }
$S = 0$
$\underset{\sim}{T} \leftarrow \{0, \dots, 0\}$
$\hat{\underset{\sim}{\theta}} = (\sum_{i=1}^{p} \underset{\sim}{c}_i \underset{\sim}{c}_i')^{-1} \sum_{i=1}^{p} \underset{\sim}{c}_i X_i$

{ PROCEDURE }
FOR $t := p+1$ TO n DO
    Calculate W_t and $\hat{\sigma}_t^2$ (using current value of $\hat{\underset{\sim}{\theta}}$) according to (3.8) and (3.10).
    $S = S + W_t$
    $\underset{\sim}{T}[t] = \hat{\sigma}_t^{-1} | S |$
    IF $\underset{\sim}{T}[t] > w(\alpha; t, n)$
        THEN We reject H_0^* on the significance level α_0.

RETURN t, $\underset{\sim}{T}$
PLOT $\{(i, \underset{\sim}{T}[i]),\ i = p+1, \ldots, t\}$
STOP

ELSE We cannot reject H_0^* on significance level α_0 (for current values of $\hat{\underset{\sim}{\theta}}$ and t).

$\hat{\underset{\sim}{\theta}} \leftarrow (\sum_{i=1}^t \underset{\sim}{c}_i \underset{\sim}{c}_i')^{-1} \sum_{i=1}^t \underset{\sim}{c}_i X_i$

Remarks:

1. We can literally repeat remark 2. of Algorithm 1. if we use any of the robust estimators described in the Section 4 instead of the classical least squares one. Of course, corresponding lines in the Algorithm 2. must be appropriately changed.
2. Neither here one should not forget about the graphical presentation of the test statistics, see the PLOT line of the Algorithm 2.
3. Selected critical values $w_i(\alpha; t, n)$ for the CUSUM M test procedures on the level $\alpha = 0,05$ & $p = 2$ are:

$w(\alpha; t, n)$	n = 40	n = 70	n = 100
$w_1(\alpha; t, n)$	$3,213\sqrt{t-2}$	$3,376\sqrt{t-2}$	$3,476\sqrt{t-2}$
$w_2(\alpha; t, n)$	$5,229 + 0,308t$	$7,358 + 0,230t$	$9,002 + 0,192t$

Like the CUSUM test, the MOSUM (moving sum) test procedures are based the recursive residuals (3.8). The corresponding critical region is of the form

$$\bigcup_{t=p+G+1}^{n} \left\{ \frac{\sigma}{\hat{\sigma}_k \sqrt{G}} \left| \sum_{k=t-G+1}^{t} W_k \right| > m(\alpha; G, n) \right\}, \tag{3.14}$$

where (3.10) can be used again as an estimate of σ^2 and $m(\alpha; G, n)$ is chosen so that the level of the test is α. For more details and another approaches see,e. g., Hackl (1980).

With the exact critical values we have the same problems as in the CUSUM case, because also here different (asymptotic) arguments lead to the different sets of critical values. Here we shall show only an approximation via the Bonferroni inequality, which gives the critical value

$$m_1(\alpha; G, n) = \sqrt{G}\, t_{1-\alpha_2}(n-p), \tag{3.15}$$

where $\alpha_2 = \dfrac{\alpha}{2(n-G-p)}$ and G is suitably chosen (it should be small relatively to n; e.g., $0,2n$ or $0,1n$). Selected critical values $m(\alpha; G, n)$ for $\alpha = 0,05$ & $p = 2$ are

G	n = 40	n = 70	n = 100
10	3,124	3,332	3,447
15	3,066	3,307	3,431
20	2,992	3,279	3,414

Let us illustrate the MOSUM method in the form of a simple algorithm, in which $\underset{\sim}{T}$ is the vector (of the length n) used for storing values of the test statistics corresponding to the respective steps of the procedure and $\underset{\sim}{P}$ and $\underset{\sim}{Q}$ are vector (of the length n) used for temporary purposes.

Algorithm 3. Classical MOSUM procedure.

{ SET UP }
$S := 0$
$\underset{\sim}{T} \leftarrow \{0, \ldots, 0\}$
$\underset{\sim}{P} \leftarrow \{0, \ldots, 0\}$
$\underset{\sim}{Q} \leftarrow \{0, \ldots, 0\}$
FOR $t := p + 1$ TO $p + G - 1$ DO
    Calculate $\hat{\underset{\sim}{\theta}}, W_t$ and $\hat{\sigma}_t^2$ according to (2.3), (3.8) and (3.10).
    $S = S + W_t$
    $\underset{\sim}{P}[t] := W_t$
    $\underset{\sim}{Q}[t] := \hat{\sigma}_t^2$

{ PROCEDURE }
FOR $t := p + G$ TO n DO
    Calculate $\hat{\underset{\sim}{\theta}}, W_t$ and $\hat{\sigma}_t^2$ according to (2.3), (3.8) and (3.10).
    $P[t] := W_t$
    $Q[t] := \hat{\sigma}_t^2$
    $S = S + W_t - P[t - G]$
    $\underset{\sim}{T}[t] = \hat{\sigma}_t^{-1} G^{-1/2} |S|$
    IF $\underset{\sim}{T}[t] > m(\alpha; G, t, n)$

THEN We reject H_0^* on the significance level α_0.
RETURN $\underset{\sim}{T}$, t
PLOT $\{(i, \underset{\sim}{T}[i]),\ i = p+G,\dots,t\}$
PLOT $\{(i, \underset{\sim}{Q}[i]),\ i = p+G,\dots,t\}$
STOP

ELSE We cannot reject H_0^* on the significance level α_0 (for current values of $\hat{\underset{\sim}{\theta}}$ and t).

Remark:

1. We can literarily repeat the remarks concerning the previous algorithms. Of course, the corresponding formulas must be appropriately changed.

4. M – ESTIMATORS AND THEIR MODIFICATIONS

Consider the linear model

$$X_i = \underset{\sim}{c}_i'\,\underset{\sim}{\theta}_0 + e_i, \quad i = 1,\dots,n, \tag{4.1}$$

where $\underset{\sim}{c}_i = (c_{i1},\dots,c_{ip})'$ are known regression vectors, $\underset{\sim}{\theta}_0 = (\theta_{01},\dots,\theta_{0p})'$ is an unknown vector parameter and $e_1,\dots,e_n$ are i.i.d. random errors with c.d.f. F fulfilling certain regularity conditions.

The *M – estimator* $\hat{\underset{\sim}{\theta}}_n(\psi)$ of $\underset{\sim}{\theta}_0$ based on observations $X_1,\dots,X_n$ and generated by a score function ψ is defined as a solution of the system of equations

$$\sum_{i=1}^{n} \underset{\sim}{c}_i\,\psi(X_i - \underset{\sim}{c}_i'\,\underset{\sim}{\theta}) = 0, \tag{4.2}$$

where the function ψ usually satisfies some mild conditions, e.g., monotonicity, certain type of smoothness etc. However, in the practice we often work with one of the following types, i.e.,

$$\psi(x) = x, \quad x \in R_1, \tag{4.3}$$

which leads to the usual *least squares estimator*,

$$\psi(x) = sign(x).\min(|x|, A), \quad x \in R_1, \tag{4.4}$$

the so called *Huber's* ψ – function,

$$\psi(x) = \begin{cases} B.sin(x/B), & |x| \le \pi B, \\ 0 & |x| > \pi B, \end{cases} \tag{4.5}$$

so called *Andrew's* ψ – function or with the *Cauchy's* ψ – function of the form

$$\psi(x) = \frac{x}{1+(x/C)^2}, \quad x \in R_1. \tag{4.6}$$

Typically one puts $A = 1,7$, $B = 1,4$ and $C = 2,4$.

Since the M – estimators are not scale invariant, modifications were developed, called studentized M – estimators. The *studentized M – estimator* $\hat{\underset{\sim}{\theta}}_n^S(\psi)$ of $\underset{\sim}{\theta}_0$ based on observations $X_1, \ldots, X_n$ and generated by a score function ψ is defined as a solution of the system of equations

$$\sum_{i=1}^{n} \underset{\sim}{c}_i \psi\left(\frac{X_i - \underset{\sim}{c}_i' \underset{\sim}{\theta}}{s_n}\right) = 0, \tag{4.7}$$

where $s_n = s_n(\underset{\sim}{X}_n)$ is an estimator of a scale functional $\sigma = \sigma(F) > 0$ with the properties

$$s_n(\underset{\sim}{X}_n) > 0, \quad s_n(a\underset{\sim}{X}_n + b(\underset{\sim}{c}_1, \ldots, \underset{\sim}{c}_n)') = a s_n(\underset{\sim}{X}_n),$$

where

$$\underset{\sim}{X}_n = (X_1, \ldots, X_n)', \quad a > 0,\ b \in R_1.$$

As for the computation of these estimators, some iterative procedure is recommended or one can use so called *one step M – estimator* $\hat{\underset{\sim}{\theta}}_n^{OS}(\psi)$ defined as

$$\hat{\underset{\sim}{\theta}}_n^{OS}(\psi) = \tag{4.8}$$

$$= \begin{cases} \hat{\underset{\sim}{\theta}}_n^* + \dfrac{1}{\gamma_n^*(\psi)} \underset{\sim}{C}_n^{-1} \displaystyle\sum_{i=1}^{n} \underset{\sim}{c}_i \psi(X_i - \underset{\sim}{c}_i' \hat{\underset{\sim}{\theta}}_n^*), & \text{if} \quad \gamma_n^*(\psi) \neq 0, \\ \hat{\underset{\sim}{\theta}}_n^*, & \text{if} \quad \gamma_n^*(\psi) = 0, \end{cases}$$

where $\hat{\underset{\sim}{\theta}}_n^*$ is a preliminary estimator fulfilling $\underset{\sim}{C}_n^{1/2}(\hat{\underset{\sim}{\theta}}_n^* - \underset{\sim}{\theta}_0) = O_p(1)$ ($\underset{\sim}{C}_n$ being defined in (2.4)) and $\gamma_n^*(\psi)$ being a consistent estimator of the derivative of $\lambda(t)$ at $t = 0$ with

$$\lambda(t) = -\int_{R_1} \psi(x - t)\, dF(x). \tag{4.9}$$

Usually one uses

$$(4.10)\qquad \gamma_n^*(\psi) =$$

$$= \frac{1}{2t\sqrt{n}} \sum_{i=1}^{n} \left[\psi\left(X_i - \underline{c}_i' \hat{\underline{\theta}}_n^* + \frac{t}{\sqrt{n}} \right) - \psi\left(X_i - \underline{c}_i' \hat{\underline{\theta}}_n^* - \frac{t}{\sqrt{n}} \right) \right],$$

where $t > 0$ is a fixed constant. The studentized version of $\hat{\underline{\theta}}_n^{OS}(\psi)$ can be introduced easily as in (4.7), i.e., replacing $(X_i - \underline{c}_i' \hat{\underline{\theta}}_n^*)$ by $(X_i - \underline{c}_i' \hat{\underline{\theta}}_n^*)/s_n$ in (4.8) and $\left(X_i - \underline{c}_i' \hat{\underline{\theta}}_n^* \pm \frac{t}{\sqrt{n}} \right)$ by $\left(X_i - \underline{c}_i' \hat{\underline{\theta}}_n^* \pm \frac{t}{\sqrt{n}} \right)/s_n$ in (4.9). For more detailed information on M-estimators see, e.g., monographs of Huber (1981) or Hampel (1985).

For our further purposes we will introduce two types of estimators useful in recursive procedures. The first one is the *recursive M-estimator* $\hat{\underline{\theta}}_k^R(\psi)$ defined as follows,

$$(4.11)\qquad \hat{\underline{\theta}}_{k+1}^R(\psi) =$$

$$= \hat{\underline{\theta}}_k^R(\psi) + (\gamma_k^R)^{-1} \underline{C}_{k+1}^{-1} \underline{c}_{k+1} \psi\left(X_{k+1} - \underline{c}_{k+1}' \hat{\underline{\theta}}_k^R(\psi) \right), \quad k \geq k_0,$$

where $\hat{\underline{\theta}}_{k_0}^R(\psi)$ is an initial estimator based on $X_1, \ldots, X_{k_0}$ (usually the M-estimator based on $X_1, \ldots, X_{k_0}$, and generated by the score function ψ) and $\gamma_k^R(\psi)$ is an estimator of the derivative of $\lambda(t)$ (cf (4.9)) at $t = 0$. An example is

$$\gamma_k^R(\psi) = \begin{cases} a_k, & \gamma_k^+(\psi) \leq a_k \\ \gamma_k^+(\psi), & a_k < \gamma_k^+(\psi) < a_k^{-1}, \qquad k \geq k_0, \\ a_k^{-1}, & \gamma_k^+(\psi) \geq a_k^{-1}, \end{cases}$$

where

$$\gamma_k^+(\psi) =$$

$$= \frac{1}{2tk} \sum_{i=1}^{k} \left[\psi\left(X_i - \underline{c}_i' \hat{\underline{\theta}}_{i-1}^R(\psi) + \frac{t}{i^\alpha} \right) - \psi\left(X_i - \underline{c}_i' \hat{\underline{\theta}}_{i-1}^R - \frac{t}{i^\alpha} \right) \right] i^\alpha,$$

with $0 < \alpha < \frac{1}{2}$, $t > 0$ fixed and $a_k \searrow 0$ for $k \to \infty$. For the properties of these estimators see Tsypkin and Poljak (1979).

The second estimator is so called *stochastic type estimator* $\hat{\underset{\sim}{\theta}}_k^{SA}(\psi)$ defined as

$$\hat{\underset{\sim}{\theta}}_{k+1}^{SA}(\psi) = \hat{\underset{\sim}{\theta}}_k^{SA}(\psi) + \frac{1}{\gamma_k^{SA}} \underset{\sim}{C}_{k+1}^{-1} \sum_{i=1}^{k+1} \underset{\sim}{c}_i' \psi\left(X_i - \underset{\sim}{c}_i' \hat{\underset{\sim}{\theta}}_k^{SA}(\psi)\right), \quad k \geq k_0,$$

where $\hat{\underset{\sim}{\theta}}_{k_0}^{SA}(\psi)$ is the initial estimator based on $X_1, \ldots, X_{k_0}$ (usually the M-estimator based on $X_1, \ldots, X_{k_0}$, and generated by the score function ψ), and $\gamma_k^{SA}(\psi)$ is an estimator of the derivative of $\lambda(t)$ at $t = 0$ which can be defined as

$$\gamma_n^{SA}(\psi) =$$

$$\frac{1}{2t\sqrt{k}} \sum_{i=1}^{k} \left[\psi\left(X_i - \underset{\sim}{c}_i' \hat{\underset{\sim}{\theta}}_k^{SA}(\psi) + \frac{t}{\sqrt{k}}\right) - \psi\left(X_i - \underset{\sim}{c}_i' \hat{\underset{\sim}{\theta}}_k^{SA}(\psi) - \frac{t}{\sqrt{k}}\right)\right]$$

with $t > 0$. The properties of this type of estimators are studied by Hušková (1991 b). Studentized versions of both $\hat{\underset{\sim}{\theta}}_k^{R}(\psi)$ and $\hat{\underset{\sim}{\theta}}_k^{SA}(\psi)$ can be defined accordingly.

5. NONRECURSIVE M – TESTS

These tests are related to the maximum likelihood ratio test statistics introduced in Section 2. The test statistics are of the form

$$\max_{p<k<n-k} Z_{nk}, \tag{5.1}$$

where Z_{nk} is one of the following statistics:

$$Z_{nk}^{(1)} = (\hat{\underset{\sim}{\theta}}_k(\psi) - \hat{\underset{\sim}{\theta}}_n(\psi))' \underset{\sim}{C}_k \underset{\sim}{C}_n^{-1} \underset{\sim}{C}_k (\hat{\underset{\sim}{\theta}}_k(\psi) - \hat{\underset{\sim}{\theta}}_n(\psi)) \hat{\sigma}_n^{-2}(\psi), \tag{5.2}$$

$$Z_{nk}^{(2)} = \underset{\sim}{S}_k'(\hat{\underset{\sim}{\theta}}_n(\psi), \psi) \underset{\sim}{C}_n^{-1} \underset{\sim}{S}_k(\hat{\underset{\sim}{\theta}}_n(\psi), \psi) \hat{\sigma}_n^{-2}(\psi) \tag{5.3}$$

and

$$Z_{nk}^{(3)} = \underset{\sim}{S}_k'(\hat{\underset{\sim}{\theta}}_n(\psi), \psi)(\underset{\sim}{C}_k^{-1} + (\underset{\sim}{C}_n - \underset{\sim}{C}_k)^{-1}) \underset{\sim}{S}_k(\hat{\underset{\sim}{\theta}}_n(\psi), \psi) \hat{\sigma}_n^{-2}(\psi). \tag{5.4}$$

Here $\hat{\underset{\sim}{\theta}}_k(\psi)$, $k = k_0, \ldots, n$, $k_0 > p$, is the M-estimator (4.2) based on $X_1, \ldots, X_k$ and generated by the score function ψ, $\underset{\sim}{C}_k$ being defined by (2.4),

$$\underset{\sim}{S}_k(\hat{\underset{\sim}{\theta}}_n(\psi), \psi) = \sum_{i=1}^{k} \underset{\sim}{c}_i \psi(X_i - \underset{\sim}{c}_i' \hat{\underset{\sim}{\theta}}_n(\psi)) \tag{5.5}$$

and

$$\hat{\sigma}_n^2(\psi) = \frac{1}{n}\sum_{i=1}^{n} \psi^2(X_i - \underset{\sim}{c}_i \underset{\sim}{\hat{\theta}}_n(\psi)) - \overline{\psi}_n^2, \tag{5.6}$$

where

$$\overline{\psi}_n = \frac{1}{n}\sum_{i=1}^{n} \psi(X_i - \underset{\sim}{c}_i' \underset{\sim}{\hat{\theta}}_n(\psi)). \tag{5.7}$$

This means that $\hat{\sigma}_n^2(\psi)$ is the estimator of $\int \psi^2(x)\, dF(x)$ (we assume that $\int \psi(x)\, dF(x) = 0$). It can be shown that

$$\underset{\sim}{S}_k'(\underset{\sim}{\hat{\theta}}_n(\psi), \psi)(\underset{\sim}{C}_k^{-1} + (\underset{\sim}{C}_n - \underset{\sim}{C}_k)^{-1})\underset{\sim}{S}_k(\underset{\sim}{\hat{\theta}}_n(\psi), \psi)\hat{\sigma}_n^{-2}(\psi)$$

has for $k \to \infty$ & $n \to \infty$ the$\chi^2$with $p-$degrees of freedom and, moreover, that $Z_{nk}^{(1)}$ and $Z_{nk}^{(2)}$ lead (asymptotically) to the same test procedures, while $Z_{nk}^{(3)}$ is different.

Sen (1984) showed that if mild conditions are satisfied, then under H_0

$$\mathcal{L}\left(\max_{p<k<n-p} Z_{nk}^{(i)}\right) \to \mathcal{L}\left(\sup_{t\in(0,1)} \sum_{j=1}^{p} B_j^2(t)\right) \quad \text{as} \quad n \to \infty, \quad i = 1, 2, \tag{5.8}$$

where $\{B_j(t),\ t \in (0,1),\ j = 1, \dots, p\}$, are independent Brownian bridges. This leads to the critical region (with asymptotic level α)

$$\max_{p<k<n-p} Z_{nk}^{(i)} > b_\alpha(p), \qquad i = 1, 2, \quad 0 < k < n - p, \tag{5.9}$$

where $b_\alpha(p)$ is the $\alpha-$quantile of the distribution of $\sup_{t\in(0,1)} \sum_{j=1}^{p} B_j^2(t)$. Tables of $b_\alpha(p)$ are in Kiefer (1959).

As for $Z_{nk}^{(3)}$, one can simply use the Bonferroni inequality which leads to the critical region (with asymptotic level α)

$$(5.10) \qquad \bigcup_{k=p+1}^{n-p} \Big\{ \underset{\sim}{S}_k'(\hat{\theta}_n(\psi),\psi)\,(\underset{\sim}{C}_k^{-1} + (\underset{\sim}{C}_n - \underset{\sim}{C}_k)^{-1}) \cdot$$

$$\cdot \underset{\sim}{S}_k(\hat{\underset{\sim}{\theta}}_n(\psi),\psi)\hat{\sigma}_n^{-2}(\psi) > \chi^2_{\alpha_3}(p) \Big\},$$

where $0 < t_0 < t_1 < 1$ are fixed, $[a]$ denotes the integer part of a, $\chi^2_{\alpha_3}(p)$ is the $(1-\alpha_3)$-quantile of the χ^2 distribution with p degrees of freedom and $\alpha_3 = \dfrac{\alpha}{2(n-p)}$.

Making use of the results of James et al. (1987) together with asymptotic arguments, one receives the critical region (with asymptotic level α)

$$(5.11) \qquad \max_{[nt_0]\le k\le[nt_1]} \Big\{ \underset{\sim}{S}_k'(\hat{\theta}_n(\psi),\psi)\,(\underset{\sim}{C}_k^{-1} + (\underset{\sim}{C}_n - \underset{\sim}{C}_k)^{-1}) \cdot$$

$$\cdot \underset{\sim}{S}_k(\hat{\underset{\sim}{\theta}}_n(\psi),\psi)\hat{\sigma}_n^{-2}(\psi) \Big\} > d_{\alpha,p}(t_0,t_1),$$

where $0 < t_0 < t_1 < 1$ and $d_{\alpha,p}(t_0,t_1)$ is the solution of the equation (2.7).

6. RECURSIVE M-TESTS

They are based on the M-recursive residuals of the form

$$(6.1) \qquad W_i(\psi) = \psi(X_i - \underset{\sim}{c}_i'\hat{\theta}_{i-1}(\psi)), \qquad i = p+1,\ldots,n,$$

where $\hat{\underset{\sim}{\theta}}_{i-1}(\psi)$ stays for any of the estimators introduced in Section 4. and based on $X_1,\ldots,X_{i-1}$. Notice that $\underset{\sim}{c}_i'\hat{\theta}_{i-1}(\psi)$ is the forecast for the i-th observation based on $X_1,\ldots,X_{i-1}$.

Similarly to the classical approach there are in principle two types of the recursive M-tests, i.e.,

1) CUSUM M-tests, based on properly standardized partial sums

$$\sum_{i=p+1}^{k} W_i(\psi), \qquad i = p+1,\ldots,n;$$

2) MOSUM M-tests, based on properly standardized moving sums

$$\sum_{i=k-G+1}^{k} W_i(\psi), \qquad i = p+G+1, \ldots, n.$$

The critical regions of the CUSUM M-test procedures are of the form

$$\bigcup_{k=k_0}^{n} \left\{ \frac{1}{\hat{\hat{\sigma}}_k}(\psi) \left| \sum_{i=p+1}^{k} W_i(\psi) \right| > w^*(\alpha; k, n) \right\}, \tag{6.2}$$

where $w^*(\alpha; k, n)$ is chosen such that the asymptotic level is α (or $\leq \alpha$), k_0 should not be too small and $\hat{\hat{\sigma}}_k^2(\psi)$ is an estimator of $\int \psi^2(x)\, dF(x)$. Here one can take, e.g.,

$$\hat{\hat{\sigma}}_k^2(\psi) = \frac{k-1}{k} \hat{\hat{\sigma}}_{k-1}^2(\psi) + \frac{1}{k} \psi^2 \left(X_k - \underset{\sim}{c}_k' \hat{\underset{\sim}{\theta}}_{k-1}(\psi) \right). \tag{6.3}$$

The crititical values $w^*(\alpha; k, n)$ are not uniquely determined. Different asymptotic arguments lead to different sets of critical values. The Bonferroni inequality leads to

$$w_1^*(\alpha; k, n) = \sqrt{k}\, \Phi^{-1} \left(1 - \frac{\alpha}{2(n-k_0)} \right), \tag{6.4}$$

where $\Phi^{-1}(t)$ denotes the quantile function corresponding to the standard normal distribution.

An application of modified results of Darling and Erdös (1956) gives

$$w_2^*(\alpha; k, n) = \tag{6.5}$$

$$= \sqrt{k} \left[\frac{-\log\log \frac{1}{\sqrt{1-\alpha}} + 2\log\log n + \frac{1}{2}\log\log\log n - \frac{1}{2}\log 4\pi}{\sqrt{2\log\log n}} \right].$$

For n large enough $w_1^*(\alpha; k, n) > w_2^*(\alpha; k, n)$, $k_0 \leq k \leq n$. Sen (1984) proposed the critical value

$$w_3^*(\alpha; k, n) = \sqrt{n}\, w_\alpha, \tag{6.6}$$

where w_α is defined by

$$P\left(\sup_{t \in [0,1]} |W(t)| > w_\alpha \right) = \alpha$$

with $\{W(t),\ t \in [0,1]\}$ being the standardized Wiener process. The critical values $w_2(\alpha; k, n)$ defined by (3.14) can be used also in this situation.

The critical regions of the MOSUM M-tests are of the form

$$\bigcup_{k=k_0}^{n} \left\{ \frac{1}{\sigma_k^*(\psi)\sqrt{G}} \left| \sum_{i=k-G+1}^{k} W_i(\psi) \right| > m^*(\alpha; h, n) \right\}, \tag{6.7}$$

$W_i(\psi)$ being defined by (6.1), $k_0 \le i \le n$, $p+1 \le k_0$ is suitably chosen, $\sigma_k^{*2}(\psi)$ being a consistent estimator of $\int \psi^2(x)\, dF(x)$, $h = G/n$ and $m(\alpha; h, n)$ being determined in such a way that the asymptotic level is α .

The Bonferroni inequality gives the critical value

$$m_1^*(\alpha; h, n) = \Phi^{-1}\left(1 - \frac{\alpha}{2(n - nh - k_0)}\right). \tag{6.8}$$

The test is asymptotically conservative.

A modification of the results of Deheuvels and Révész (1987) leads, for $h = G/n$ small, to the critical value

$$m_2^*(\alpha; h, n) = \sqrt{2 \log \frac{1}{h}} + \frac{\log\log \frac{1}{h} + \log \frac{4}{\pi} - 2 \log\log \frac{1}{1-\alpha}}{\sqrt{8 \log \frac{1}{h}}}. \tag{6.9}$$

For n large enough $m_1^*(\alpha; h, n) > m_2^*(\alpha; h, n)$.

One should note that all presented critical values $w_i^*(\alpha; k, n)$ and $m_i^*(\alpha; h, n)$ are based on the asymptotic distributions; hence n should be large enough. Further, k_0 should not be chosen too small because both the estimators $\hat{\theta}_k(\psi)$ and $\hat{\sigma}_k^2(\psi)$ can considerably oscillate for small k. Theoretical results are derived by Sen (1984) and Hušková (1989).

The CUSUM M-test and the MOSUM M-test are not scale invariant (similarly as the usual M-estimators) which means that critical regions corresponding to the observations $X_1, \ldots, X_n$ and $cX_1, \ldots, cX_n$, $0 < c \ne 1$, are generally different. However, studentization of the described procedures gives scale invariant procedures. The studentization consists in replacing

$$X_i - \underset{\sim}{c}_i' \hat{\theta}_{i-1}(\psi) \quad \text{and} \quad X_i - \underset{\sim}{c}_i' \hat{\theta}_k(\psi)$$

by

$$\frac{X_i - \underset{\sim}{c}_i'\hat{\theta}_{i-1}(\psi)}{s_k} \quad \text{and} \quad \frac{X_i - \underset{\sim}{c}_i'\hat{\theta}_k(\psi)}{s_k}.$$

Here s_k is an estimator of a scale functional $\sigma = \sigma(F) > 0$ with the properties

$$s_k(\underset{\sim}{X}_k) > 0, \quad s_k(a\underset{\sim}{X}_k + b(\underset{\sim}{c}_1, \dots, \underset{\sim}{c}_k)' = as_k(\underset{\sim}{X}_k),$$

where $\underset{\sim}{X} = (X_1, \dots, X_k)'$, $a > 0$, $b \in R_1$. Since s_k can considerably oscillate for small k, $\sigma(F)$ should preferably be estimated independently of $X_1, \dots, X_n$.

7. Estimation of the change point τ

Since the change point τ is usually estimated by $t_{\widehat{m}}$, where $\widehat{m}$ is an estimator of m, it suffices to estimate m. All test statistics considered in Sections 2,3,5 and 6 can be used to construct an estimator of m.

For simplicity we shall focus here on a simple model

$$X_i = \theta_1 + (\theta_n - \theta_1)\mathrm{I}_{\{i>m\}} + e_i, \qquad i = 1, \dots, n, \tag{7.1}$$

where θ_1, θ_n and m are unknown parameters, $m/n \doteq \gamma \in (0,1)$ and $e_1, \dots, e_n$ are i.i.d. random errors with distribution $N(0,1)$.

The maximum likelihood estimator $\widehat{m}_1$ of m is defined as

$$\widehat{m}_1 = \underset{1 \le k < n}{\arg\max} \left\{ \frac{nk}{n-k} \left(\overline{X}_k - \overline{X}_n\right)^2 \right\}.$$

The CUSUM method (see Section 3.A.) leads to the estimator

$$\widehat{m}_2 = \underset{1 < t \le n}{\arg\min} \left\{ \left| \sum_{k=2}^{t} W_k \right| > w(\alpha; t, n) \right\},$$

where $w(\alpha; t, n)$ is described below the formula (3.2) with α very small.

The MOSUM method (see Section 3.A.) leads to the estimator

$$\widehat{m}_3 = \underset{G+2 \le t \le n}{\arg\max} \left\{ \left| \sum_{k=t-G+1}^{t} W_k \right| \right\} - G.$$

All these estimators are consistent if $\sqrt{n}(\theta_n - \theta_1) \to \infty$ as $n \to \infty$ and can be used to construct confidence intervals for m. Further information on the point estimation and the confidence intervals for m can be found, e. g., in Cobb (1978) or Siegmund (1988).

Robust estimators of m can be introduced in either of the following ways:

$$\widehat{m}_4(\psi) = \arg\max_{1\le k\le n}\left\{\left|\sum_{i=1}^{k}\psi\left(X_i - \widehat{\theta}_n(\psi)\right)\right|\right\}$$

or

$$\widehat{m}_5(\psi) = \arg\min_{k_0<k\le n}\left\{\sum_{i=p+1}^{k}\left|\left(\psi(X_i - \widehat{\theta}_{i-1}(\psi))\right)\right|\sigma_k^{*-1}(\psi) > \right.$$

$$\left. > w^*(\alpha; k, n)\right\}$$

or

$$\widehat{m}_6 = \arg\max_{G+2\le k\le n}\left\{\left|\sum_{i=k-G+1}^{k}\psi(X_i - \hat{\theta}_{i-1}(\psi))\right| G^{-1/2}\,\sigma_k^{*-1}(\psi) > \right.$$

$$\left. > m^*(\alpha; G/n, n)\right\} - G,$$

where $\hat{\theta}_{i-1}(\psi)$ is the $M-$estimator defined as a solution of the equation

$$\sum_{j=1}^{i-1}\psi(X_i - \theta) = 0$$

and all other symbols were introduced in Sections 5 and 6.

Under mild conditions these estimators are consistent (random errors need not be normally distributed). Simulations made by the authors show reasonable behavior of all these estimators, however, $\widehat{m}_5(\psi)$ seems to behave slightly better than $\widehat{m}_4(\psi)$ and $\widehat{m}_3(\psi)$. If the distribution of the errors is nonnormal, than the estimators $\widehat{m}_j(\psi)$, j=4,5,6, behave better than $\widehat{m}_1, \widehat{m}_2, \widehat{m}_3$.

On some other robust estimators of m one can learn in the recent papers by Carlstein (1988), Ritov (1990) and Dümbengen (1991).

ACKNOWLEDGEMENT

The authors wish to thank to Professor Hackl and Professor Michálek for carefull reading the manuscript and valuable remarks.

References

[1] Antoch J., *Recursive M-tests for the change point problem: Monte Carlo study*, submitted (1991).

[2] Antoch J. and Hušková M., *Some M-tests for detection of a change in linear models*, Proceedings of the 4th Prague Symposium on Asymptotic Statistics (Hušková M. and Mandl P., eds.), Charles University Press, Praha, 1989, pp. 123-136.

[3] Bagshaw M. and Johnson R. A., *Sequential procedures for detecting parameter change in a time series models*, Journal of American Statistical Association **72** (1977), 593-597.

[4] Basseville M. and Benveniste A., *Detection of abrupt changes in signals and dynamical systems*, Lecture Notes in Control and Information Sciences, vol. 77 (Basseville M. et al., eds.), Springer-Verlag, New York, 1986.

[5] Bhattacharya G. K. and Johnson R. A., *Nonparametric tests for shift at an unknown time point*, Annals of Mathematical Statistics **39** (1968), 1731-1743.

[6] Broemling L. D. and Tsurumi H., *Econometrics and Structural Change*, Serie Statistics, vol. 74, M. Dekker, New York, 1986.

[7] Brown R. L., Durbin J. and Evans J. M., *Techniques for testing the constancy of regression relationships over time (with discussion)*, Journal of Royal Statistical Society **B37** (1975), 149-192.

[8] Carlstein E., *Nonparametric change-point estimation*, Annals of Statistics **16** (1988), 188-197.

[9] Chernoff H. and Zacks S., *Estimating the current mean of a normal distribution which is subject to changes in time*, Annals of Mathematical Statistics **35** (1964), 999-1018.

[10] Chin Choy J. H. and Broemling L. D., *Some Bayesian inferences for a changing linear model*, Technometrics **22** (1980), 71-78.

[11] Cobb G. W., *The problem of Nile conditional solution to a change point problem*, Biometrika **65** (1978), 243-251.

[12] Csörgö M. and Horváth L., *Nonparametric tests for the change point problem*, Journal of Statististical Planning and Inference **17** (1987), 1-9.

[13] Csörgö M. and Horváth L., *Nonparametric methods for the change point problem*, Handbook of Statistics, vol. 7 (P.R. Krishnaiah and C.R. Rao, eds.), J. Wiley, New York, 1988, pp. 403-425.

[14] Deheuvels P. and Révész P., *Weak laws for the increments of Wiener processes, Brownian bridges, empirical processes and partial sums of i.i.d. r.v.'s.*, Mathematical Statistics and Probability Theory, vol. A (Puri M.L. et al., eds.), D. Reidel, Berlin, 1987, pp. 69-88.

[15] De Long D., *Crossing probabilities for a square root boundary by a Bessel process*, Communications in Statistics-Theory Methodology **A10** (1981), 2197-2213.

[16] Deshayes J. and Picard D., *Off-line statistical analysis of change point models using nonparametric and likelihood methods*, Lecture Notes in Control and Information Sciences, vol. 77 (Basseville M. et al., eds.), Springer-Verlag, New York, 1986, pp. 103-168.

[17] Dienske H. and Metz J. A. J., *Mother-infant body contact in macaques: A time interval analysis*, Biology of Behaviour **2** (1977), 3-37.

[18] Dübengen L., *The asymptotic behavior of some nonparametric change-point estimators*, Annals of Statistics **19** (1991), 1471-1475.

[19] Esterby S. R. and El Shaarawi A. H., *Inference about the point of change in a regression model*, Applied Statistics **30** (1981), 277 - 285.

[20] Farley J. U. and Hinich M. J., *A test for a shifting slope coefficient in a linear model*, Journal of American Statistical Association **65** (1970), 1320 - 1329.

[21] Ferreira P. E., *A Bayesian analysis of a switching regression model - known number of regimes*, Journal of American Statistical Association **70** (1975), 370 - 374.

[22] Gardner L. A. Jr., *On detecting changes in the mean of normal variates*, Annals of Mathematical Statistics **40** (1969), 116 - 126.

[23] Hackl P., *Testing the constancy of regression relationships over time*, Vandenhoeck and Ruprecht, Göttingen.

[24] Hackl P. and Westlund A., *Statistical analysis of structural change: An annotated bibliography*, Empirical Economics **14** (1989), 167 - 172.

[25] Hackl P. (ed.), *Statistical Analysis and Forecasting of Economic Structural Change*, Springer - Verlag, New York, 1989.

[26] Haccou P., Dienske E. and Meelis E., *Analysis of time inhomogenity in Markov chains applied to mother - infant interactions of rhesus monkeys*, Animal Behaviour **31** (1983), 927 - 945.

[27] Haccou P., Meelis E. and van der Geer S., *On the likelihood ratio test for a change point in a sequence of independent exponentially distributed random variables*, Report MS - R 8507, CWI, Amsterdam, 1985.

[28] Hampel F. R., Ronchetti E. M., Rousseeuw P. J. and Stahel W. R., *Robust Statistics: The Approach Based on Influence Functions*, J. Wiley, New York, 1985.

[29] Hawkins D. M., *A note on continuous and discontinuous segmented regression*, Technometrics **22** (1980), 443 - 444.

[30] Hawkins D. L., *Sequential procedures for detecting deviations in the parameters of the autoregressive model from specified targets*, Sequential Analysis **3** (1984), 121 - 154.

[31] Hawkins D. L., *Sequential procedures for monitoring the parameters of the autoregressive model relative to unspecified targets*, Sequential Analysis **4** (1985), 275 - 310.

[32] Hawkins D. L., *Retrospective and sequential tests for a change in distribution based on Kolmogorov - Smirnov type statistics*, Sequential Analysis **7** (1988), 23 - 51.

[33] Hinkley D. V., *Time - ordered classification*, Biometrika **59** (1972), 509 - 523.

[34] Hsu D. A., *Detecting shifts of parameter in gamma sequences with applications to stock price and air traffic flow analysis*, Journal of American Statistical Association **74** (1979), 31 - 40.

[35] Huber P. J., *Robust Statistics*, J. Wiley, New York, 1981.

[36] Hušková M., *Recursive M - tests for detection of change*, Sequential Analysis **7** (1988 a), 75 - 90.

[37] Hušková M., *Adaptive procedures for detection of change*, Statistics and Decisions **6** (1988 b), 137 - 148.

[38] Hušková M., *Nonparametric tests for change in regression models at an unknown time point*, International Workshop on Theory and Practice in Data Analysis, Report R - MATH - 01/89, Akademie der Wissenschaften der DDR, Berlin, 1989 a, pp. 101 - 124.

[39] Hušková M., *Some asymptotic results for robust procedures for testing the constancy of regression models over time*, Kybernetika **26** (1990 a), 392 - 403.

[40] Hušková M., *Asymptotics for robust MOSUM*, Commentationes Mathematicæ Universitatis Carolinæ **31** (1990b), 345 – 356.

[41] Hušková M., *Recursive M – tests for the change point problem*, Economic Structural Change, Analysis and Forecasting (Hackl P. and Westlund A.H., eds.), IIASA & Springer – Verlag, New York, 1991 a, pp. 13 – 34.

[42] Hušková M., *Stochastic approximation type estimators in linear models*, Sequential Analysis **10** (1991), 45-68.

[43] Hušková M. and Sen. P. K., *Nonparametric tests for shift and change in regression at an unknown time point*, Statistical Analysis and Forecasting of Economic Structural Change (Hackl P., ed.), Springer – Verlag, New York, 1989, pp. 71 – 85.

[44] James B., James K. L. and Siegmund D., *Tests for a change – point*, Biometrika **74** (1987), 71 – 84.

[45] James B., James K. L. and Siegmund D., *Conditional boundary crossing probabilities with applications to change point problems*, Annals of Probability **16** (1988), 825 – 839.

[46] Johnson R. A. and Bagshaw M., *The effect of serial correlation on the performance of CUSUM tests*, Technometrics **16** (1974), 103 – 112.

[47] Kiefer J., *K – sample analogues of the Kolmogorov – Smirnov and Cramér – von Mises tests*, Annals of Mathematical Statistics **30** (1959), 420 – 447.

[48] Kim Hyune – ju and Siegmund D., *The likelihood ratio test for a change point in simple linear regression*, Biometrika **76** (1979), 409 – 423.

[49] Krishnaiah P. R. and Miao B. Q., *Review about estimation of change points*, Handbook of Statistics, vol. 7 (Krishnaiah P.R. and Rao R.C., eds.), Elsevier, Amsterdam, 1988, pp. 375 – 402.

[50] Lombard F., *Rank tests for change point problem*, Biometrika **74** (1987), 615 – 624.

[51] Page E. S., *A test for a change in a parameter occuring at an unknown time point*, Biometrika **42** (1955), 523 – 527.

[52] Page E. S., *On problems in which a change in a parameter occurs at an unknown point*, Biometrika **44** (1957), 248 – 252.

[53] Parzen E., *Comparison change analysis*, Manuscript of the lecture in International Symposium on Nonparametric Statistics and Related Topics, Carleton University Press, Carleton, 1991.

[54] Pettitt A. N., *A nonparametric approach to the change point problem*, Applied Statistics **28** (1979), 126 – 135.

[55] Picard O., *Testing and estimating change points in time series*, Advances in Applied Probability **17** (1985), 841 – 867.

[56] Poljak B. T. and Tsypkin Ya. Z., *Adaptive estimation algorithms*, Automation and Remote Control **3** (1979), 72 – 84.

[57] Praagman J., *Bahadur efficiency of rank tests for the change point problem*, Annals of Statististics **16** (1988), 198 – 217.

[58] Ritov Y., *Asymptotic efficient estimation of the change – point with unknown distributions*, Annals of Statistics **18** (1990), 1829 – 1839.

[59] Sastri T., Flores B. and Valdéz J., *Detecting points of change in time series*, Computation and Operational Research **16** (1988), 271 – 293.

[60] Schechtman E. and Wolfe D. A., *Distribution – free tests for the change point problem*, Technical Report, Ohio State University Press, Ohio, 1981.

[61] Schulze U., *Mehrphasenregression: Stabilitätsprüfung, Schätzung, Hypothesenprüfung*, Akademie Verlag, Berlin, 1986.
[62] Segen J. and Sanderson A. C., *Detecting change in a time series*, IEEE transactions on information theory **26** (1980), 249 - 255.
[63] Sen A. and Srivastava M. S., *On tests for detecting change in mean*, Annals of Statistics **3** (1975), 98 - 108.
[64] Sen P. K., *Asymptotic theory of some tests for a possible change in the regression slope occuring at an unknown time point*, Zeitschrift für Wahrscheinlichkeitsteorie und verw. Gebiete **52** (1980), 203 - 218.
[65] Sen P. K., *Asymptotic theory of some tests for constancy of regression relationships over time*, Statistics **13** (1982), 21 - 31.
[66] Sen P. K., *On some recursive residual rank tests for change points*, Recent Advances in Statistics. Papers in Honor of Herman Chernoff's Sixtieth Birthday (Rizvi M.H., ed.), Academic Press, New York, 1983a, pp. 371 - 391.
[67] Sen P. K., *Test for change points based on recursive U-statistics*, Sequential Analysis **1** (1983b), 263 - 284.
[68] Sen P. K., *Recursive M-tests for the constancy of multivariate regression relationships over time*, Sequential analysis **3** (1984), 191 - 211.
[69] Shaban S. A., *Change point problem and two phase regression: An annotated bibliography*, International Statistical Review **48** (1980), 83 - 93.
[70] Shiryayev A. N., *On optimum methods in quickest test detection problem*, Theory of Probability and its Applications **10** (1963), 22 - 46.
[71] Shiryayev A. N., *Optimal Stopping Rules*, Springer - Verlag, New York, 1978.
[72] Siegmund D., *Boundary crossing probabilities and statistical applications*, Annals of Statististics **14** (1986), 361 - 404.
[73] Siegmund D., *Confidence sets in change - point problems*, International Statistical Review **56** (1988), 31 - 38.
[74] Wolfe D. A. and Schechtman E., *Nonparametric statistical procedures for the change point problem*, Journal of Statistical Planning and Inference **9** (1984), 389 - 396.
[75] Worsley K. J., *An improved Bonferroni inequality and applications*, Biometrika **69** (1982), 297–302.
[76] Worsley K. J., *Testing for a two - phase multiple regression*, Technometrics **25** (1983), 35 - 42.
[77] Worsley K. J., *Confidence regions and tests for a change - point in a sequence of exponential family variables*, Biometrika **73** (1986), 91 - 104.
[78] Zacks S., *Survey of classical and Bayesian approaches to the change point problem: Fixed sample and sequential procedures of testing and estimation*, Recent Advances in Statistics. Papers in Honor of Herman Chernoff's Sixtieth Birthday (Rizvi M.H., ed.), Academic Press, New York, 1983, pp. 245 - 269.
[79] Zacks S., *Detection and Change - point Problem*, Handbook of Sequential Analysis (Ghosh B.K. and Sen P.K., eds.), Serie Statistics, vol. 118, M. Dekker, New York, 1991.

Charles University, Department of Probability and Mathematical Statistics, Sokolovská 83, CS - 186 00 Praha 8, Czechoslovakia; tel: (+422) 231 78 62; fax: (+422) 84 76 88; telex: 121 673 mff.

E-mail: ANTOCH@CSPGUK11.BITNET, HUSKOVA@CSPGUK11.BITNET

ROBUST ESTIMATION IN LINEAR MODEL AND ITS COMPUTATIONAL ASPECTS

JAROMÍR ANTOCH, JAN ÁMOS VÍŠEK

Charles University, Prague
Czechoslovak Academy of Sciences, Prague

ABSTRACT. Heuristics of robust estimation is briefly explained and a survey of the most typical robust methods of the point estimation both of the location and scale parameters and in the linear model is presented. Basic notions of robust statistics as well as of some particular topics of robust diagnostics are discussed, too. Testing of submodels and the possibilities of adaptive approach in the estimation of parameters in the linear model are also commented.

1. ROBUSTNESS AND ITS PHILOSOPHY

Most of the results of the classical parametric statistics is based on, and derived under, some (usually rather "narrow") assumptions about the underlying probabilistic model. Typical example is a solution of the regression problem, which was achieved by the least squares (LS) method under an assumption that the distribution of errors in the model is normal. Under this assumption the estimator is known to be unbiased with minimal variance among all estimators. Sometimes it was clear from the obtained residuals that the assumption about normality is questionable. But for a routine application it was argued that even without normality LS estimator has minimal variance among all linear estimators. Besides this, it was generally accepted for a long time in the case of the regression problem that although precise normality can

1990 *Mathematics Subject Classification.* 62F35, 62F10, 62F12, 62J05, 62H15.

Key words and phrases. Robust estimation and testing, location and scale parameters, linear model, $M-$, $L-$, $R-$, $W-$, $P-$ and $S-$estimators, minimal distance estimators, adaptive estimators, one- and $k-$step estimators, invariance, consistency, influence function, gross-error sensitivity, small-shift sensitivity, rejection point, break-down point, asymptotic representation, asymptotic normality, tail behavior, stability of models, testing differences of models.

hardly be assumed to hold, the variance of the LS-estimator is near the smallest possible. Moreover, it has been commonly believed that small deviations from the assumptions could cause only small changes in the behavior of statistical procedures.

Unfortunately, it turned out that this belief has been totally wrong. Despite the fact that examples demonstrating that believe are transparently simple and well known, a lot of applied statisticians still unconsciously rely on some kind of a "continuous behavior" of statistical procedures when "continuously deviating" from idealized assumptions. But in any arbitrarily small neighborhood of the normal distribution we can find a distribution which totaly "breaks" the least squares method. This is a theoretical phenomenon, but unfortunately it has a real correspondence significant from the practical point of view, i. e., the contamination of the data. Therefore, to safeguard us against it we should solve corresponding probabilistic problems under the assumption that the data are generated by a probability measure from an appropriately determined family of measures. *And that is a corner-stone idea of robust statistics, i. e., that the data should be modeled by a sufficiently rich family of measures and a desired solution would take into account all such distributions.* For example, we may try to find an estimator which minimizes, in a given family of estimators, maximum of variances and is as "efficient" as possible. However, it is not the main goal of robust statistics to achieve a maximum possible efficiency.

It has been "proved" by the practice, at least at the beginning of a data analysis, that it is much more important to take measures against bad, and sometimes even fatal, influence of "wrong" observations. The word "wrong" includes not only observations which evidently do not belong to the model, but also observations which are "atypical" for the model, i. e., those which can occur with very small probability. In any case it is important to notice that the latter case may have the same destructive influence on the results of statistical procedures as the former one.

So, we may summarize our goals as follows:

- We would like to have procedures which have a reasonable efficiency at the "central" model usually assumed in the classical statistics and which are moreover "robust" in the sense that small deviations from this model should cause only mild, or if you like, insignificant, deviations of results.
- Further, and it is also an important requirement, that substantial deviations from the assumed model can change results somewhat more, but should not imply a catastrophe, i. e., the total break-

down of the procedures. The exact sense will be specified in Section 5.

According to the historical development, the main areas into which activities in robust statistics can be divided are the following ones:

- Description of the bulk of the data by a parametric model.
- Identification of and warning about (highly) influential points, i.e., the points which (probably significantly) deviate from the bulk of the data. The attention has been first payed to the search for *outliers* and only somewhat later the attempts were directed to the detection of *leverage points*, too.
- Construction of the procedures which are able to cope with the influential points, preferably without rejecting them.
- Solution of the problem of deviations from an expected correlation structure, or more generally, dependence–independence structure.

Let us present a simple example describing the behavior of two classical statistical procedures which will illustrate the situation. The example is due to Tukey, but has been also used in Huber (1977), e.g. Let us assume that we have n independent random variables $X_1, \ldots, X_n$ identically distributed (i.i.d. r.v.'s) according to

$$F_\varepsilon(x) = (1-\varepsilon)\Phi\left(\frac{x-\mu}{\sigma}\right) + \varepsilon\Phi\left(\frac{x-\mu}{3\sigma}\right), \qquad \mu \in R_1,\ \sigma > 0,$$

where Φ denotes the standard normal distribution function (d.f.) and ε, $0 \le \varepsilon \le 1$, is the so called contamination level, see Section 2 for more details. Let us consider two estimators of the scale parameter, i.e.,

$$s_n = \sqrt{\frac{1}{n}\sum_{i=1}^{n}(X_i - \overline{X}_n)^2}$$

and

$$d_n = \sqrt{\frac{\pi}{2}} \cdot \frac{1}{n}\sum_{i=1}^{n}\left|X_i - \overline{X}_n\right|.$$

Please notice that the role of the factor $\sqrt{\pi/2}$ is to remove the bias of d_n in the normal case. Then the asymptotic relative efficiency (ARE) of s_n and d_n, which is understood here as the ratio of the asymptotic variances of s_n and d_n, is equal to

ε	0	0.001	0.002	0.005	0.010	0.020	0.050
ARE	0.876	0.948	1.016	1.198	1.439	1.752	2.035

This demonstrates that only 2 bad observations in the sample of the size 1 000 may completely destroy the better behavior of s_n with respect to d_n for purely normal case. Moreover, taking into account that present opinion of applied statisticians is that in the practice one finds contamination levels ϵ somewhere between 0.01 and 0.1 with upper bound attained much more frequently than it is commonly anticipated, we would prefer the use of d_n to s_n in practice. Interesting discussion can be found in Stigler (1977, 1986), e.g.

On the other hand, if we assume that $X_1, \ldots, X_n$ are i.i.d. r.v.'s distributed according to the Student's t_ν, then the relative efficiency (RE) of s_n and d_n, which is understood here as the ratio of the variances of s_n and d_n, is equal to

ν	3	4	5	6	7	8	10	12	∞
RE	1.253	1.129	1.086	1.064	1.051	1.043	1.032	1.026	1

This shows that also in the case of the random sample from the Student's distribution t_ν we should prefer the use of d_n to s_n.

At the end of this section we would like repeat once more that robust statistics is not a part of the nonparametric statistics. On the contrary, parametric models of the classical statistics are substituted by, more or less complicated, families of distributions in such a way that around every member of a parametric system we build up a neighborhood. This may be a convex combination of the Huber type $\mathcal{F}_\epsilon$, Kolmogorov – Smirnov or Prochorov distance based etc. In other words, the parametric model is imbedded into a "neighborhood of the model".

Throughout the paper N will denote the set of all integers, R_1 the real line and R_p p-dimensional Euclidean space. Let us also remark that all vectors will be understood to be column-wise and denoted by subtilde.

2. Model and its possible violations.

In this section we will concentrate on the question how "small violations" of the statistical model, which will be represented by means of a family of probability distributions

$$\mathcal{F} = \{F_\theta \mid \theta \in \Theta \subseteq R_p,\ \Theta \text{ open and convex set}\},$$

can look like. A typical example of the parametric model will be as follows, i.e., $\Omega = R_1$, $\Theta = R_1$ and F_θ ($\equiv \Phi_\theta$) the standard normal distribution with *theta* $= (\mu, 1)$.

The first typical situation, which is quite frequent in practice, is a presence of outlying observations (outliers) among the data. On one side the outliers are real errors due to disturbation (of any type) and should be excluded from the data set, of course, after careful checking that it is really so. The situation is much more difficult in the hypothetic case of the participant of our Summer School with the weight 260 kg. On one hand it is a true value and probably the case interesting for physicians. On the other hand it violates quite seriously usually assumed normal model and the question arises what to do with it.

From the mathematical point of view we have several possibilities how to cope with such a situation. One of the most natural approaches which we will follow in this section, consists in the idea that we observe the mixture of "good" and "bad" observations. While the "good" observations can be described by means of the normal model Φ_θ, e.g., we can assume that the "bad" observations follow another model generated by some distribution H. Such a situation may be described easily by the distributions from the model

$$(1) \qquad \mathcal{F}_\varepsilon = \{F \,|\, F = (1-\varepsilon)\Phi_\theta + \varepsilon H,\ H \text{ any distribution}\},$$

where ε, $0 \le \varepsilon \le 1$, establishes the ratio of the "bad" observations (at least approximately). Model (1) is one of the most frequently used alternatives to the normal model $\mathcal{F}_0$ and is usually called *ε-contaminated normal model.* For H one uses, as a rule, the distribution with heavier tails than those of the normal one forming the "true" model. This means the distribution for which corresponding density converges to zero more slowly than the density of the normal one. Notice that the contamination need not be symmetric. Nevertheless, it should be pointed out that the nonsymmetric contamination can cause the bias of the resulting procedures. For more detailed discussion see Collins (1976), Hampel et al. (1986) or Huber (1964, 1981) among others.

Let us go back to our example with the hypothetic participant of our Summer School with an outlying weight. It is very important to take into account that it is necessary to distinguish very carefully between the outliers from the point of view of the statistician and his client, the physician, e.g., because this is not always the same. While the weight 260 kg of the participant of our Summer School is from the statistician's point of view a typical outlier violating his otherwise "nice" normal model and should be excluded from the data set, from the point of view of medicine it may be a very interesting case which should be followed carefully. Therefore, the conclusion is that automatic trimming off the outlying data or their weighting down, respectively, is not the best solution in all cases. On the contrary, such data should be seriously reconsidered and only then one should decide what to do with them, i.e., whether to use another model, trim them off, check these cases once more etc.

Another typical operation is rounding off the data or grouping them into the classes. If we have (again) a good reason for the description of "theoretical" data set by the normal model Φ_θ, it is advantageous to use for their "rounded off" (or "grouped") version a distribution from the set

$$\mathcal{K}_\delta = \left\{ F \;\middle|\; \sup_{x \in R_1} |\Phi_\theta(x) - F(x)| \le \delta \right\}, \tag{2}$$

where $\Phi_\theta(x)$ denotes normal distribution with the mean θ and variance 1, while δ, $0 \le \delta \le 1$, establishes maximal possible distance between the hypothetical and the true distribution function. Model (2) is in literature usually called Kolmogorov – Smirnov neighborhood of the normal model.

It is evident that if the values of ε and δ are small enough, both model (1) and model (2) represent only small (negligible) violation of the respective model. In the first case there can appear arbitrarily large values, but only in very few cases. In the second case the opposite is true. We violate the assumed normal model for many observations, but what is important, "only a little".

It is clear that there can be violated also some other assumptions one usually uses, e.g., that of the independence of the observations (errors) etc. Nevertheless, we will drop this problem here and we refer to Martin and Yohai (1986), Hampel (1986) or Franke et al. (1984) instead.

3. Functional representation of the estimators.

Let $T : \mathcal{F}_T \longmapsto R_1$ be a functional defined on a set of probability distributions $\mathcal{F}_T$ and F_n, $n = 1, 2, \ldots$, denotes empirical distribution function corresponding to the random sample $X_1, \ldots, X_n$ from the distribution $F_\theta \in \mathcal{F}_T$. Assume moreover, that the functional T is Fisher consistent, i.e., $T(F_\theta) = \theta \;\; \forall \theta \in \Theta$. Then, one can use for an estimation of θ sequence of real statistics $\{T_n,\; n = 1, 2, \ldots\}$, where

$$T_n = T(F_n) = T(X_1, \ldots, X_n), \qquad n = 1, 2, \ldots . \tag{3}$$

Examples of estimators defined in this way will be shown in the sequel. More general definitions can be found in literature, e.g., estimators which can be represented by the functionals only asymptotically etc. For more details see Fernholz (1983), Hampel (1986), Huber (1981) or Serfling (1980) among others.

One of the advantages of such a functional representation consists of the fact, that using general theorems describing the distribution of statistical functionals we can prove (relatively easily) asymptotic normality of underlying estimators. This means that we can show, under relatively mild conditions, that

$$\mathcal{L}_F\left(\sqrt{n}(T_n - T(F))\right) \longrightarrow \mathcal{N}(0, \mathsf{V}(T,F)), \qquad n \to \infty, \tag{4}$$

where $\mathcal{L}_F(\dots)$ represents the distribution of ... under F and $\mathsf{V}(T,F)$ is called the asymptotic variance of the sequence $\{T_n,\ n = 1,2,\dots\}$ given by (3). For details see Fernholz (1983) or Serfling (1980).

4. Influence function.

It is well known from practice that the behavior of the sample mean can be totally destroyed even by only one observation. More precisely, if the value of this additional observation will go to infinity, then the value of the sample mean will proceed in the same way. This is a typical example how only one "bad" observation can cause disaster, i.e., the total break down of the procedure. Notice on the contrary that this is not the case of $\alpha-$ trimmed mean, sample median, Huber's $M-$ estimator and many other ones.

As it has been emphasized earlier, the functional representation of the estimators allow us to introduce many of their important characteristics. The key role among them is played by the so called *influence function*, which shows among others the way leading to the solution of the following problems:

"What will happen if we add to the very large sample $X_1, \dots, X_n$ an additional observation with an arbitrary value and how can we measure the influence of this new observation on the resulting estimator?"

The influence function of the functional T at the point F is defined as

$$\mathsf{IF}(x; T, F) = \lim_{\varepsilon \searrow 0} \frac{T((1-\varepsilon)F + \varepsilon\Delta_x) - T(F)}{\varepsilon}, \qquad x \in R_1, \tag{5}$$

where Δ_x is a probability measure which puts mass 1 to the point x. In other words, the influence function describes the effect of an infinitesimal contamination at the point x on the estimator, standardized by the mass of the contamination. The definition goes back to Hampel (1968, 1971).

If $\mathsf{IF}(x;\ T, F)$ is unbounded, as is the case of the sample mean for example, then it is clear that even only one outlying observation (outlier) can cause catastrophe (failure, break down) of the underlying estimator, which means here that the estimator can attain the value $\pm\infty$. We will illustrate forms of IF's for some of the most typical estimators of location later on.

Aside this basic use of the influence function there exists a series of other applications starting from definition of measures of robustness up to the possibility to establish (asymptotical) efficiency of the estimators. Most of them are more or less deeply connected with the following important property of the influence function. If we are interested in exact asymptotic variance $\mathsf{V}(T, F)$ in (4), it is possible to show that

$$\mathsf{V}(T,F) = \int_{R_1} (\mathsf{IF}(x;\ T,F))^2\ dF(x), \tag{6}$$

of course, under the fulfillment of certain regularity conditions, see Huber (1981) or Serfling (1980) for details. It is evident that for many commonly used estimators one can find $\mathsf{V}(T, F)$ directly and in much more easier way than through (6). On the other hand, it is important to notice that (6) gives a correct answer in almost all practical situations. This is especially important when trying to establish the relative asymptotic efficiency of any pair of estimators defined as the ratio of their asymptotic variances.

Another interesting property of the influence function is the possibility to establish the asymptotic Rao – Cramer inequality for the sequence of Fisher consistent estimators $\{T_n,\ n = 1, 2, \dots\}$. More precisely, if $0 < \mathsf{J}(F_\theta) < \infty$, where $\mathsf{J}(F_\theta) = \mathsf{E}_{F_\theta}\,(\partial \ln f_\theta(x)/\partial\theta)^2$ and $f_\theta(x)$ is density corresponding to $F_\theta(x)$, then

$$\int_{R_1} (\mathsf{IF}(x;\ T,\ F_\theta))^2\ dF_\theta(x) \geq \frac{1}{\mathsf{J}(F_\theta)}, \qquad \forall\, \theta \in \Theta,$$

where equality holds if and only if $\mathsf{IF}(x;\ T, F_\theta)$ is proportional to

$$\partial \ln f_\theta(x)/\partial\theta.$$

5. Measures of robustness.

There appears often the question *how to measure robustness*. From different measures suggested in literature we will discuss those four which can be used for measuring of robustness for the following types of model violation:

- the existence of the gross errors (outliers);
- rounding off the data;
- grouping the observations into the classes;
- the fact that statistical model is only an (better or worse) idealization of the reality.

Let us consider a sequence of estimators $\{T_n,\ n = 1, 2, \dots\}$ defined through some functional T (see (3)) and let F be some fixed probability distribution, i.e., the model. One of the approaches how to measure the sensitivity of an estimator to the existence of gross errors is the so called *gross error sensitivity measure* γ^* defined as

$$(7) \qquad \gamma^* = \gamma^*(T, F) = \sup_x \ |\,\mathsf{IF}(x;\, T, F)\,|\,.$$

Of course, one takes the supremum in (7) only over those $x \in R_1$ for which the influence function is defined (exists). Roughly speaking, γ^* gives the worst possible influence which a small contamination by gross errors can have on the value of the estimator.

From the point of the robustness of estimators it is desirable to have γ^* finite, because in such a case a new additional observation with arbitrary large value cannot destroy them. Estimators with γ^* finite are usually called *B – robust estimators*.

A close connection to the previous measure has the so called *rejection point* ϱ^*, defined for symmetric distribution F as

$$(8) \qquad \varrho^* = \varrho^*(T, F) = \inf_{r>0} \ \{\mathsf{IF}(x;\, T, F) = 0 \quad \text{for} \quad |x| > r\}\,.$$

One puts $\varrho^* = \infty$ if there does not exist such a r. It is clear that if ϱ^* is finite, then contamination by a new observation in the region of $\{x \,|\, \mathsf{IF}(x;\, T, F) = 0\}$ cannot influence the estimator. This is closely connected with the fact that all observations further away than ϱ^* are rejected completely. In other words, measure (8) is more than closely connected with the old statistical practice to trim off extreme observations in the sample, say the maximum and the minimum.

When rounding off the data, or grouping them into the classes, their values change only a little. However, what is important is the fact that this concerns most of them. Intuitively, one can measure the effect of the shift of one observation from the point x to the point y by means of the value $\mathsf{IF}(y;\, T, F) - \mathsf{IF}(x;\, T, F)$. Nevertheless, we are usually interested in the worst possible effect in practice. This can be measured, e.g., by means of the *local shift sensitivity measure* λ^* defined as

$$\lambda^* = \lambda^*(T, F) = \sup_{x \neq y} \frac{|\mathsf{IF}(y; T, F) - \mathsf{IF}(x; T, F)|}{|y - x|}. \tag{9}$$

From the mathematical point of view expression (9) is nothing else than the smallest Lipschitz constant of the order one for the influence function. Notice, that the value $\lambda^* = \infty$ is typical not only for highly nonrobust estimators, but also in the case when $\mathsf{IF}(x;\, T, F)$ has at least one jump. This is the case of a sample median for example. Generally speaking, the importance of λ^* is less than that of γ^* or ϱ^*. On the other hand, one should use the estimators with $\lambda^* < \infty$ and thus better than the sample median.

The measures of robustness γ^*, ϱ^* and λ^* just introduced can be considered as local measures, because the influence function plays a crucial role in their definition. On the other hand one needs also some global measures of robustness to be able to cope with the problem of influence of more than one observation on the result of his/her procedure. From many possibilities suggested in literature we will concentrate on the so called finite sample breakdown point. For some other measures of this type and comparison of them we refer to Donoho and Huber (1983), e.g. To simplify the notation we will consider only the case of the estimators of location and shift parameters.

Let $X_1, \ldots, X_n$ be a random sample from the distribution $F(x - \beta)$, where β is an unknown location parameter and $x_1, \ldots, x_n$ is the realization of this random sample, i.e., the observed values. Then the *finite sample breakdown point* ε_n^* of the estimator T_n of β is defined as

$$\begin{aligned} \varepsilon_n^* &= \varepsilon_n^*(T_n;\, x_1, \ldots, x_n) = \\ &= \frac{1}{n} \max \left\{ m \,\middle|\, \max_{i_1, \ldots, i_m} \sup_{y_1, \ldots, y_m} |T_n(z_1, \ldots, z_n)| < \infty \right\}, \end{aligned} \tag{10}$$

where the sample $\{z_1, \ldots, z_n\}$ is obtained by replacing any m values $x_{i_1}, \ldots, x_{i_m}$ by the arbitrary values $y_1, \ldots, y_m$. In other words, the finite sample break down point ε_n^* shows how many values we can change without breaking down the procedure, i.e., without causing a catastrophe. Here the catastrophe is equivalent again to the fact that T_n will attain the value $\pm\infty$. It is interesting to notice that ε_n^* usually does not depend on the realization $(x_1, \ldots, x_n)$ of $(X_1, \ldots, X_n)$, and only little on the sample size n.

The finite sample breakdown point has been defined in (10) for estimators of the location parameter only and it is important to notice that for other types of estimators its definition must be properly modified. We will illustrate such a modification by the estimator of the scale parameter. Let $X_1, \ldots, X_n$ be random sample from the distribution $F(x/\sigma)$, $\sigma > 0$ be an unknown scale parameter, and $(x_1, \ldots, x_n)$ is the realization of this random sample. Let T_n^* be any estimator of σ. Then, the finite sample breakdown point of T_n^* is defined again by (10), however, under the additional condition

$$\min_{i_1,\ldots,i_m} \inf_{y_1,\ldots,y_m} T_n^*(z_1, \ldots, z_n) > 0.$$

The reason for introduction of this last condition is clear, i.e., to avoid nonpositive values of the estimators of scale.

At the end of this section we would like to introduce the last measure of the robustness we will refer to. It is the so *called asymptotic breakdown point* ε^* defined as

$$\varepsilon^* = \lim_{n\to\infty} \varepsilon_n^*. \tag{11}$$

6. Robust estimators of the location and scale parameters.

In this section we will concentrate on the construction of robust estimators of location and scale parameters. If we take a look at the literature about robustness, we can see that the greatest attention has been payed just to this case and especially to the estimators of the location parameter. The reason is not only that this setup is very simple and allows elegant solutions, but especially because the corresponding solutions show the way how to proceed in many, much more general, situations, e.g., in the study of linear models etc. Our attention will be concentrated especially on $M-$ $L-$ and $R-$estimators, but some other classes as $P-$, $W-$, $S-$ and $D-$estimators will be mentioned as well. The main reason is that most of them are now not only of the theoretical interest, but they can either be found in packages such as BMDP, S+, SAS and SPSS, or easily programmed by the user.

6a. *L*– estimators of the location parameter.

Let $X_1, \ldots, X_n$ be a random sample from the distribution $F(x - \beta)$ with unknown location parameter β and let $X_{(1)} < \cdots < X_{(n)}$ be the corresponding order statistics. Then, we understand by the L–estimator of β any statistic of the form

$$\widehat{L}_n = \sum_{i=1}^{n} c_i X_{(i)}, \tag{12}$$

where c_i, $i = 1, \ldots, n$, $\sum_{i=1}^{n} c_i = 1$, are known constants generated by a weight function $h : [0,1] \longmapsto R_1$, $\int_0^1 h(x)\,dx = H \neq 0$, and given by

$$c_i = H^{-1} \int_{(i-1)/n}^{i/n} h(x)\,dx, \qquad i = 1, \ldots, n. \tag{13}$$

In practice one usually chooses nonnegative values c_i. It is evident that their choice should reflect the effort to diminish, eventually totally avoid, the possible influence of the outlying observations on the resulting value of the estimator. In other words, one usually puts smaller weights on the "extreme" order statistics than to the "central" ones to prevent the collapse of the procedure.

Both simplicity of computation and easiness of interpretation belong to the most appreciable advantages of L– estimators. The most typical examples are the sample mean and the sample median, α– trimmed and α– Winsorized means, quick and dirty estimators like the Gastwirth's one etc.

It is clear that L– estimators of location can be defined either using a weight function $h(x)$ and relations (12) – (13) or through the functional approach. Both ways are illustrated in Examples 1 and 2. The natural question how to choose the weight function $h(x)$ for given data set in practice is closely connected with them. Nevertheless, the response is rather complicated and depends in a crucial way on the optimality criterion used and some additional assumptions about the model. More details can be found in Andrews et al. (1972) or Hampel et al. (1986), e.g.

Before proceeding to the promised examples, let us stop for a moment with the asymptotic distribution of the L– estimators of location. They are, under mild and quite natural regularity conditions, asymptotically normal with the asymptotic variance

(14) $$\mathsf{V}(T,F) =$$

$$\int \left(\frac{\int_0^x h(F(y))\,dy - \int_{R_1} \left(\int_0^t h(F(y))\,dy \right) dF(t)}{\int_{R_1} h(F(y))\,dF(y)} \right)^2 dF(x).$$

Let us notice finally that if the weights c_i are of the form (13), then the $L-$estimators of location parameter are Fisher consistent.

Example 1. The most typical and most frequently used one in practice is the class of so called $\alpha-$ *trimmed means*, i.e., the class of the statistics of the form

(15) $$\widehat{L}_{n,\alpha} = \frac{1}{n-2[n\alpha]} \sum_{i=[n\alpha]+1}^{n-[n\alpha]} X_{(i)},$$

where $0 \leq \alpha < 1/2$ is a fixed constant called the *proportion of the trimming* and $[n\alpha]$ denotes the largest integer which is less than or equal to $n\alpha$. It is evident that statistics $\widehat{L}_{n,\alpha}$ form a natural generalization of the old statistical rule to throw away the minimal and maximal value from the sample and to compute the sample mean from the rest of the observations. In our case we trim off $[n\alpha]$ observations on both sides of the ordered sample. The choice of the trimming proportion α should be always slightly higher than the expected contamination by "bad" observations in practice. As a special case we will have the sample mean for $\alpha = 0$, while the sample median for $\alpha \cong 1/2$. Generalization of (15) for nonsymmetric trimming is evident. Let us remind that in literature one can find also different refinements of (15), see Andrews et al. (1972), Huber (1981) or Hampel et al. (1986) for details.

It is easy to show that the functional generating the class of $\alpha-$trimmed means is of the form

(16) $$T(F) = \frac{1}{1-2\alpha} \int_\alpha^{1-\alpha} F^{-1}(t)\,dt,$$

where $F^{-1}(t) = \sup\{x \mid F(x) \leq t\}$ denotes the inverse distribution function. To show this it is sufficient to substitute the empirical distribution function F_n corresponding to the random sample $X_1, \ldots, X_n$

into (16) instead of F. Similarly, it is easy to show that the weight function $h(x)$ for the class of α-trimmed means is of the form

$$h(x) = I_{[\alpha,1-\alpha]}(x), \qquad x \in [0,1], \tag{17}$$

where $I_A(x)$ denotes the indicator function.

The influence function for the α-trimmed mean is of the form

$$\mathsf{IF}(x;\,T,F) = \tag{18}$$

$$\begin{cases} (F^{-1}(\alpha) - c(\alpha))/(1-2\alpha), & x < F^{-1}(\alpha), \\ (x - c(\alpha))/(1-2\alpha), & F^{-1}(\alpha) \le x \le F^{-1}(1-\alpha), \\ (F^{-1}(1-\alpha) - c(\alpha))/(1-2\alpha), & x > F^{-1}(1-\alpha), \end{cases}$$

where

$$c(\alpha) = \frac{1}{1-2\alpha}\int_{\alpha}^{1-\alpha} F^{-1}(t)\,dt + \alpha[F^{-1}(\alpha) + F^{-1}(1-\alpha)].$$

It follows immediately from (18) that the influence function for the sample mean is of the form

$$\mathsf{IF}(x;\,T,F) = \begin{cases} x - \int_0^1 F^{-1}(t)\,dt, & F \text{ nonsymmetric}, \\ x, & F \text{ symmetric}. \end{cases} \tag{19}$$

Moreover, we can see directly from (18) that for the α-trimmed mean

$$\gamma^* = \max\left\{\left|F^{-1}(\alpha)\right|, \left|F^{-1}(1-\alpha)\right|\right\},$$

$\lambda^* = 1$ and $\varepsilon_n^* = [\alpha n]$, i.e., the representants of this class of estimators are B-robust for $\alpha > 0$ and the asymptotic breakdown point ε^* is equal to the trimming proportion α.

Example 2. The second most important class of the L-estimators of the location parameter β is formed by the weighted sum of several order statistics chosen according to some "well justified rule", i.e., the estimators of the form

$$\widehat{L}_n = \sum_{j=1}^{k} c_{i_j} X_{(i_j)}, \tag{20}$$

where $1 \le i_1 < \cdots < i_k \le n$, $1 \le k \le n$ and $\sum_{j=1}^{k} c_{i_j} = 1$.

Typical examples are the sample median and the so called *Gastwirth's estimator* $\widehat{L}_{nG}$ defined as

(21) $$\widehat{L}_{nG} = 0.3X_{([n/3])} + 0.4X_{([n/2])} + 0.3X_{([2n/3])}.$$

We will appreciate them especially when we have an ordered sample available, and need to find a fast orientation in it.

It is easy to show that estimators of the form (20) can be generated by means of the functionals of the form

(22) $$T_k(F) = \sum_{j=1}^{k} c_{i_j} F^{-1}\left(t_{i_j}\right),$$

or by the weight function $h = \sum_{j=1}^{k} c_{i_j} \delta_{t_{i_j}}$, respectively, where δ_x is the probability measure which puts mass 1 to the point x, $0 < t_{i_1} < \cdots < t_{i_k} < 1$, $1 \le k \le n$ and $\sum_{j=1}^{k} c_{i_j} = 1$.

Due to the fact that the influence function for a general p–quantile q_p is of the form

(23) $$\mathsf{IF}(x; q_p, F) = \begin{cases} \dfrac{p-1}{f(F^{-1}(p))}, & x < F^{-1}(p), \\ \dfrac{p}{f(F^{-1}(p))}, & x > F^{-1}(p), \end{cases}$$

we obtain immediately the form of the influence function of $\widehat{L}_n$. In fact, it is the function with jumps of the size $c_{i_j}/f(F^{-1}(t_{i_j}))$ at points $F^{-1}(t_{i_j})$ and constant elsewhere.

Let us concentrate on the most typical representant of this class, i.e., on the sample median. It follows immediately from (23) that its influence function is of the form

$$\mathsf{IF}(x; T, F) = \frac{1}{2f(F^{-1}(1/2))} \operatorname{sign}\left(x - F^{-1}\left(\frac{1}{2}\right)\right), \qquad x \in R_1.$$

Moreover, it is easy to show that this estimator is B–robust and attains the asymptotic breakdown point 1/2. On the other hand, the corresponding value of the local shift sensitivity measure $\lambda^* = +\infty$. Despite the fact that this type of nonrobustness is not so crucial, it follows from here that we still should look for better estimators than the sample median is and should use them.

According to our expectation we can show similar results also for the Gastwirth's estimator $\widehat{L}_{nG}$, which is B–robust, $\varepsilon_n^* = 1/3$ and $\lambda^* = +\infty$.

6b. L–estimators of the scale parameter.

Let $X_1, \ldots, X_n$ be random sample from the distribution $F(x/\sigma)$ with unknown scale parameter σ and let $X_{(1)} < \cdots < X_{(n)}$ be the corresponding order statistics. Then we understand by the L–estimator of σ any statistic of the form (12) again, but the weights are defined now with the help of a weight function $g : [0,1] \longmapsto R_1$, $\int_0^1 g(t)F^{-1}\,dt \neq 0$, as

$$c_i = \frac{\displaystyle\int_{(i-1)/n}^{i/n} g(t)\,dt}{\displaystyle\int_0^1 g(t)\,F^{-1}(t)\,dt}. \tag{24}$$

If the underlying distribution function $F(x)$ is assumed to be symmetric, the weights are usually standardized so that $\sum_{i=1}^n c_i = 0$. This last condition can be easily achieved using a skew symmetric function g, i.e., to use $g(x)$ such that $g(x) = -g(1-x)\ \ \forall x \in [0,1]$.

Typical representants of L–estimators of scale parameter σ consist of the class of α–*interquantile deviations* (called sometimes also α–*quantile range*) generated by the weight function $g(t) = \delta_{1-\alpha} - \delta_\alpha$, $0 < \alpha < 1/2$, where δ_p denotes the delta function. The *interquartile deviation* corresponding to the choice of $\alpha = 1/4$ is the best known among them. They are B–robust and $\varepsilon^* = \alpha$.

6c. R–estimators of the location parameter.

Let $\underset{\sim}{X} = (X_1, \ldots, X_n)$ be a random sample from the (symmetric) distribution $F(x-\beta)$ with an unknown parameter β. R–estimators of β can be defined in several different ways. Nevertheless, the common feature of all these approaches is the fact that they are closely connected to the rank tests of hypotheses about β. In this paragraph we will use the approach based on the sign rank test of the hypothesis $\beta = \beta_0$. It is well known that such a test can be based on the statistic

$$\begin{aligned} S^+ &= S^+(\underset{\sim}{X} - \beta_0) = S^+(X_1 - \beta_0, \ldots, X_n - \beta_0) = \\ &= \sum_{i=1}^n \operatorname{sign}\,(X_i - \beta_0)\,a\left(\frac{R_i^+(\beta_0)}{n+1}\right), \end{aligned} \tag{25}$$

where $R_i^+(\beta_0)$ is the rank of $|X_i - \beta_0|$ among $|X_1 - \beta_0|, \ldots, |X_n - \beta_0|$ and $a(\cdot)$ is a score function.

Let us assume that $a(t) = \varphi((t+1)/2)$, $0 < t < 1$, where $\varphi : [0,1] \longmapsto R_1$ is a nondecreasing skew symmetric function, i.e., $\varphi(x) = -\varphi(1-x)\ \forall\ x \in [0,1]$, such that $\int_0^1 \varphi^2(t)\,dt < \infty$. Then, the statistic $S^+(\underset{\sim}{X} - t)$ is nonincreasing in t, attains both positive and negative values with probability one and $\mathsf{E}_{\beta_0} S^+(\underset{\sim}{X} - \beta_0) = 0$. This allows to define the *R–estimator* $\widehat{R}_n$ of β as the solution of the equation

$$S^+(\underset{\sim}{X} - t) = 0. \tag{26}$$

It is evident, due to the form of the statistic S^+, that the solution of the equation (26) can be attained exactly only in very exceptional cases. One of the rare exceptions is the so called ***Hodges - Lehmann estimator***, i.e., the statistic of the form

$$\widehat{R}_{nHL} = \underset{1\leq i,j\leq n}{\text{med}} \left\{ \frac{X_i + X_j}{2} \right\},$$

which can be obtained using the score function corresponding to the Wilcoxon sign test. Therefore, most of the authors prefer to define the R–estimator of β as

$$\widehat{R}_n = \frac{1}{2}\left(\widehat{R}_n^+ + \widehat{R}_n^-\right), \tag{27}$$

where

$$\widehat{R}_n^+ = \sup\left\{ t \,\middle|\, S^+(\underset{\sim}{X} - t) \geq 0 \right\}$$

and

$$\widehat{R}_n^- = \inf\left\{ t \,\middle|\, S^+(\underset{\sim}{X} - t) \leq 0 \right\}.$$

There exists also another possibilities to define R–estimators of the location parameter based on the two sample rank tests. See Hettmansperger (1985) for details.

R–estimators of the location parameter are rather of a theoretical importance. They are especially popular among theoretical statisticians. The main reason is the fact that they could use the results derived for rank tests for the study of their properties. The use of R–estimators is only occasional in practice.

6d. R–estimators of the scale parameter.

As far as we know, R–estimators of the scale parameter were not thoroughly studied in literature.

6e. $M-$estimators.

In this paragraph we will concentrate on $M-$estimators of location and scale parameters. Unlike what we have done for the previous types of estimators, we will describe the general situation first and only then illustrate it for the special cases as the estimators of location and scale are.

Let $X_1, \ldots, X_n$ be a random sample from the distribution $F_\theta(x)$, $x \in R_1$, with unknown parameter $\theta \in \Theta \subseteq R_1$. It is natural to estimate θ via the maximum likelihood method, i.e., to use that value $\widehat{T}_n = \widehat{T}_n(X_1, \ldots, X_n)$ which maximizes the joint density

$$\prod_{i=1}^{n} f_{T_n}(X_i) \tag{28}$$

with respect to T_n, or, equivalently its logarithm. This leads to the minimization problem

$$\min_{T_n} \sum_{i=1}^{n} (-\ln f_{T_n}(X_i)) . \tag{29}$$

Let $\rho : R_1 * \Theta \longmapsto R_1$ be any real function. The natural generalization of the maximum likelihood estimator proposed by Huber (1964) consists in replacement of the minimization problem (29) by

$$\min_{M_n} \sum_{i=1}^{n} \rho(X_i; M_n) . \tag{30}$$

It is evident that if there exists[1] a function $\psi(x,\theta) = \partial/\partial\theta\rho(x,\theta)$, then the solution of (30) must satisfy the equation

$$\sum_{i=1}^{n} \psi(X_i; M_n) = 0. \tag{31}$$

On the other hand, not every solution of (31) must be equivalent to the solution of the minimization problem (30). The reason is that it can correspond to some of the local minima of $\sum_{i=1}^{n} \rho(X_i; M_n)$. Despite

[1] Agreement: By the symbol ρ (respectively ψ), we will denote the function of two variables $x \in R_1$ and $\theta \in \Theta$, as well as its special cases when θ is location or scale parameter, compare with (32) ((34) respectively).

this slight inconsistency we will define the $M-$ *estimator* of θ as any solution of the minimization problem (30) or any solution of the equation (31), respectively, and denote it $\widehat{M_n}$. Other modifications can be found in literature . For better insight and illustration, one of them will be presented later on when discussing in more details $M-$estimators of the location parameter.

The role of the $\rho-$function is to weight down the influence of potential outliers in the sample on resulting value of the estimator. One attains this when using $\rho-$functions converging to ∞ for $|x| \to +\infty$ more slowly than the tails of the function x^2, i.e., the $\rho-$function generating $M-$estimators corresponding to the least squares estimator. A summary of the most commonly used $\rho-$functions together with adequate $\psi-$functions and some additional information can be found in Table 1.

6f. $M-$estimators of the location parameter.

Let $X_1, \ldots, X_n$ be random sample from the distribution $F(x-\beta)$ where $\beta \in R_1$ is an unknown parameter. Put

$$(32) \quad \rho(x,\beta) = \rho(x-\beta) \quad \& \quad \psi(x,\beta) = \psi(x-\beta), \quad \beta \in R_1, \; x \in R_1.$$

It is common to use $\psi(x)$ antisymmetric provided that $F(x)$ is symmetric, i.e., to insist that $\forall\, x \in R_1 \;\; \psi(x) = -\psi(-x)$.

We will understand by $M-$ *estimator* $\widehat{M_n}$ of β any solution of the minimization problem (30), or any solution of equation (31), respectively, where $\rho(x,\beta)$ and $\psi(x,\beta)$ are given by (32).

It is relatively easy to find the conditions under which M-estimators of the location parameter are consistent and asymptotically normal, see Serfling (1980), e.g., for details. It follows that their asymptotic variance is of the form

$$(33) \qquad \mathsf{V}(T,F) = \int \left(\frac{\psi(x)}{\int_{R_1} \psi'(t)\, dF(t)} \right)^2 dF(x),$$

provided that there exists $\partial/\partial\beta\; \psi(x-\beta) = \psi'(x-\beta)$ with the exception of a finite number of points, and $\int_{R_1} \psi'(t)\, dF(t) \neq 0$.

It is interesting to compare (33) with (6). We can see immediately that the influence function of $M-$estimators of the location parameter

Name	$\rho(x)$	$\psi(x)$	$w(x) = \frac{\psi(x)}{x}$	range of x
Andrews (sine)	$A^2\left(1-\cos\left(\frac{x}{A}\right)\right)$ $2A^2$	$A\sin\left(\frac{x}{A}\right)$ 0	$\frac{A}{x}\sin\left(\frac{x}{A}\right)$ 0	$\lvert x\rvert \le \pi A$ $\lvert x\rvert > \pi A$
Tukey	$\frac{B^2}{6}\left(1-\left(1-\left(\frac{x}{B}\right)^2\right)^3\right)$ $\frac{B^2}{6}$	$x\left(1-\left(\frac{x}{B}\right)^2\right)^2$ 0	$\left(1-\left(\frac{x}{B}\right)^2\right)^2$ 0	$\lvert x\rvert \le B$ $\lvert x\rvert > B$
Cauchy	$\frac{C^2}{2}\ln\left(1+\left(\frac{x}{C}\right)^2\right)$	$\frac{x}{1+(x/C)^2}$	$\frac{1}{1+(x/C)^2}$	$x \in R_1$
Fair	$D^2\left(\frac{\lvert x\rvert}{D}-\ln\left(1+\frac{\lvert x\rvert}{D}\right)\right)$	$\frac{x}{1+\lvert x\rvert/D}$	$\frac{1}{1+\lvert x\rvert/D}$	$x \in R_1$
Huber	$\frac{x^2}{2}$ $E\lvert x\rvert-\frac{E^2}{2}$	x $E\,\mathrm{sign}(x)$	1 $\frac{E}{\lvert x\rvert}$	$\lvert x\rvert \le E$ $\lvert x\rvert > E$

Table 1. Several typical weight functions generating M-estimators.

Name	$\rho(x)$	$\psi(x)$	$w(x) = \frac{\psi(x)}{x}$	range of x
Logistic	$F^2 \ln\left(\cosh\left(\frac{x}{F}\right)\right)$	$F \tanh\left(\frac{x}{F}\right)$	$\frac{F}{x} \tanh\left(\frac{x}{F}\right)$	$x \in R_1$
Trimmed	$\frac{x^2}{2}$	x	1	$\lvert x \rvert \le G$
Huber	$\frac{G^2}{2}$	0	0	$\lvert x \rvert > G$
Welsh	$\frac{H^2}{2}\left(1 - \exp\left\{-\left(\frac{x}{H}\right)^2\right\}\right)$	$x \exp\left\{-\left(\frac{x}{H}\right)^2\right\}$	$\exp\left\{-\left(\frac{x}{H}\right)^2\right\}$	$x \in R_1$
L_p-norm	$\lvert x \rvert^p$	$p\lvert x \rvert^{p-1} \operatorname{sign}(x)$	$p\lvert x \rvert^{p-2}$	$x \in R_1$
Hampel	$\frac{x^2}{2}$	x	1	$\lvert x \rvert \le I$
	$I\lvert x \rvert - \frac{1}{2}I^2$	$I \operatorname{sign}(x)$	$\frac{I}{\lvert x \rvert}$	$I < \lvert x \rvert \le 1,25I$
	$\frac{12}{7} I\lvert x \rvert - \frac{2}{7}x^2 - \frac{28}{56}I^2$	$\left\{\frac{12}{7}I - \frac{4}{7}\lvert x \rvert\right\} \operatorname{sign}(x)$	$\frac{12I}{7\lvert x \rvert} - \frac{4}{7}$	$1,25I < \lvert x \rvert \le 3I$
	$\frac{116}{56} I^2$	0	0	$\lvert x \rvert > 3I$

Table 1. Several typical weight functions generating M–estimators (continuation).

is proportional to the function $\psi(x)$ generating M–estimators. More precisely,

$$\mathsf{IF}\,(x;\,T_\psi,F) = \frac{\psi(x)}{\displaystyle\int_{R_1} \psi'(t)\,dF(t)}. \tag{34}$$

This means that the influence function of M–estimators of the location parameter can be also used as the function defining M–estimators.

One of the greatest disadvantages of M–estimators of location is the fact that they are not scale invariant. Therefore, the so called studentized M–estimators have been proposed and studied. More precisely, the ***studentized M – estimator of location*** can be defined as any solution $\widehat{M}_n^S$ of the equation

$$\sum_{i=1}^{n} \psi\left(\frac{X_i - M_n}{S_n}\right) = 0, \tag{35}$$

where S_n is an estimator of the scale parameter. It is desirable to use some robust estimator also for S_n. The most common advice found in literature is to use here the statistic

$$\widehat{\sigma}_{MAD} = \frac{1}{0.6745} \operatorname*{med}_{1\le i\le n} \left\{ \left| X_i - \operatorname*{med}_{1\le j\le n} X_j \right| \right\}, \tag{36}$$

i.e., the weighted median of the deviations from the median of the sample. Most authors prefer this approach to that of estimating location and scale parameters simultaneously.

Another (rather serious) disadvantage of M–estimators of location is that for non – monotone ψ functions the solution of (31) need not be unique. There exist several ways to overcome this problem, e.g.:

- to take from all solutions of the equation (31) that one for which (30) is minimized;
- to use that solution which is closest to the apriori fixed initial estimator (sample median, interquartile mean etc.);
- to follow the same approach as for R–estimators, see (26), i.e., to put

$$\widehat{M}_n = \frac{1}{2}\left(\widehat{M}_n^+ + \widehat{M}_n^-\right), \tag{37}$$

where

$$\widehat{M}_n^+ = \sup\left\{ t \,\middle|\, \sum_{i=1}^{n} \psi(X_i - t) \ge 0 \right\}$$

and

$$\widehat{M}_n^- = \inf\left\{t \,\middle|\, \sum_{i=1}^n \psi(X_i - t) \le 0\right\}.$$

Let us go back to equation (35). It is well known that an iterative algorithm can be used for finding its solution. The simplest way is to start from some initial estimator, say $\widehat{M}_n^{(0)} = \underset{1\le i\le n}{\operatorname{med}} X_i$ and $\widehat{S}_n^{(0)}$ given by (36), and to define

$$\widehat{M}_n = \widehat{M}_n^{(0)} + \widehat{S}_n^{(0)} \frac{\displaystyle\sum_{i=1}^n \psi\left(\frac{X_i - \widehat{M}_n^{(0)}}{\widehat{S}_n^{(0)}}\right)}{\displaystyle\sum_{i=1}^n \psi'\left(\frac{X_i - \widehat{M}_n^{(0)}}{\widehat{S}_n^{(0)}}\right)}. \tag{38}$$

This last estimator is usually called in literature *one-step M-estimator* and it is easy to see, that it is nothing else than the first Newton iteration for solving equation (35). In most cases the estimators of this type have better properties than the initial one. This has been shown not only theoretically, but has been confirmed several times by extensive Monte Carlo studies. It is also evident is that we can repeat an iteration step several times, i.e., to put

$$\widehat{M}_n^{(j+1)} = \widehat{M}_n^{(j)} + \widehat{S}_n^{(0)} \frac{\displaystyle\sum_{i=1}^n \psi\left(\frac{X_i - \widehat{M}_n^{(j)}}{\widehat{S}_n^{(0)}}\right)}{\displaystyle\sum_{i=1}^n \psi'\left(\frac{X_i - \widehat{M}_n^{(j)}}{\widehat{S}_n^{(0)}}\right)}, \quad j = 0, 1, \ldots, k. \tag{39}$$

The estimators defined through (39) are called *k-step M-estimators* and are important especially for two reasons. At first, they show a way how to find the solution of equation (31). At second, they give the possibility to improve, often in quite remarkable way, the behavior of some rather poor initial estimator which may be robust but not efficient. More information about k-step M-estimators can be found in Huber (1981).

Example 3. One from the most commonly discussed ψ-functions is the so called Huber's ψ-function given by

$$\psi(x) = \begin{cases} x, & |x| \le k, \\ k\,\operatorname{sign}(x), & |x| > k, \end{cases} \tag{40}$$

where $0 < k < +\infty$ is a fixed constant. It is easy to show that the M-estimator corresponding to the ψ-function of the form (40), called in literature usually ***Huber's M-estimator***, is B-robust, $\varrho^* = +\infty$ and $\varepsilon^* = 1/2$.

Now it is well known that Huber's ψ-function is not the optimal choice for many real problems. This can be easily illustrated by examples comprising leverage points in a data set (discussion with more details will be included in the part devoted to the regression analysis). On the other hand, its "historical influence" was extremely important and till now we can find papers studying its properties and proposals of new algorithms for its evaluation etc., see Ekblom and Madsen (1989), e.g. Nevertheless, in our presentation we would like to look at Huber's M-estimator from another point of view, i.e., how it is possible to derive $\psi(x)$ of the form (40) and illustrate in this way one of the basic approaches commonly used in robust estimation.

Huber in (1964) considered a simple model of gross errors, i.e., the data which can be described using the distribution from the set

$$\mathcal{G}_\varepsilon = \{G \,|\, G = (1-\varepsilon)F_\theta + \varepsilon H\}\,, \tag{41}$$

where F_θ is any distribution function describing the (basic) model, H is any (not necessarily symmetric) distribution function describing its contamination and ε, $0 \le \varepsilon \le 1$, is a fixed constant establishing the proportion of the contamination. The special case of (41), ε-contaminated normal model, has been already considered earlier, see (1).

If we fix both the (basic) model F_θ and the function ψ generating the M-estimator, it is easy to find its asymptotic variance using (33). Huber's basic idea consists in the search for that function ψ_0 in the class Ψ of all available ψ-functions generating the M-estimator, which minimizes the maximal asymptotic variance in $\mathcal{G}_\varepsilon$. More precisely, the main aim is to find that function $\psi_0 \in \Psi$, for which

$$\sup_{G\in\mathcal{G}_\varepsilon} \mathsf{V}(T_{\psi_0}, G) = \min_{\psi\in\Psi} \sup_{G\in\mathcal{G}_\varepsilon} \mathsf{V}(T_\psi, G)\,. \tag{42}$$

Let G_0 be a distribution function minimizing Fisher's information for all $G \in \mathcal{G}_\varepsilon$ and let us denote $G_0'(x) = g_0(x)$. Then, the solution of (42) is of the form $\psi_0(x) = -g_0'(x)/g_0(x)$ and the distribution G_0 is called the ***least favorable distribution for the class*** $\mathcal{G}_\varepsilon$. It is easy to show that if one uses for the description of the data model (1), i.e.,

$\varepsilon-$contaminated normal distribution, then ψ_0 is just of the form (40) and k can be found from the equation

$$2\Phi(k) - 1 + \frac{2\phi(k)}{k} = \frac{1}{1-\varepsilon}, \tag{43}$$

where $\Phi(x)$, $\phi(x)$ respectively, is the distribution function and density of the standard normal distribution. This minimax approach has been generalized later both for some other neighborhood types and to the other types of robust estimators, see Bickel (1981) or Jurečková (1991), e.g.

6g. ***M*–estimators of the scale parameter.**

Let $X_1, \ldots, X_n$ be random sample from distribution $F(x/\sigma)$, where $x \in R_1$ and $\sigma > 0$ is an unknown scale parameter. Let us put

$$\psi(x,\sigma) = \psi\left(\frac{x}{\sigma}\right), \qquad x \in R_1,\ \sigma > 0. \tag{44}$$

Moreover, let us assume that ψ is symmetric provided that F is symmetric, too.

We will understand by M– *estimator of the scale parameter* σ any solution of equation (31), where ψ is given by (44), and denote it $\widehat{S}_n$. It is evident that the properties of $\widehat{S}_n$ will be similar to those in the location case. Under certain regularity conditions and if $\int_{R_1} t\psi'(t)\,dF(t) \neq 0$, the asymptotic normality of $M-$estimators of the scale parameter can be shown, i.e., validity of (4) with the asymptotic variance

$$\mathsf{V}(T,F) = \int \left(\frac{\psi(x)}{\int_{R_1} t\psi'(t)\,dF(t)} \right)^2 dF(x). \tag{45}$$

Moreover, if $\int_{R_1} \psi(x)\,dF(x) = 0$, then the corresponding $M-$estimator of σ is Fisher consistent.

Let us assume that $F(x)$ is symmetric and $\psi(x)$ is both symmetric and strictly monotone. Then, it is relatively easy to show that:

a) if $\psi(x)$ is moreover bounded, then the corresponding $M-$estimator is $B-$robust and $\varepsilon^* = -\psi(0)/(\psi(\infty) - \psi(0)) \leq 1/2$;
b) if $\psi(x)$ is not bounded, then the corresponding $M-$estimator is not $B-$robust and $\varepsilon_n^* = 0$ for all $n \geq 1$.

Example 4. The most commonly used $M-$estimator of the scale parameter is the MAD – estimator defined in (36). This estimator is $B-$robust and $\varepsilon^* = 1/2$.

6h. P–estimators of the location parameter.

Let $X_1, \dots, X_n$ be a random sample from the distribution $F(x - \beta)$, where $x \in R_1$ and $\beta \in R_1$ is an unknown location parameter. Then, the P–estimator of β, which we will denote by $\widehat{P}_n$, is defined by

$$\widehat{P}_n = \frac{\displaystyle\int_{R_1} \beta \prod_{i=1}^{n} \gamma(X_i - \beta)\, d\beta}{\displaystyle\int_{R_1} \prod_{i=1}^{n} \gamma(X_i - \beta)\, d\beta}, \tag{46}$$

where $\gamma(x)$ is any probability density function (not necessarily corresponding to the "true" model). The name comes from the fact that these estimators form a natural generalization of the Pitman estimator, see Johns (1979) for details.

It is easy to show that the influence function corresponding to the P–estimators is of the form

$$\mathsf{IF}(x; T, F) = \frac{\dfrac{d}{dx} \ln \gamma(x)}{\displaystyle\int_{R_1} \left(\frac{d^2}{dx^2} \ln \gamma(x) \right) dF(x)}.$$

Therefore, it follows that for each P–estimator there exists an M–estimator with the same influence function (and hence with the same asymptotic variance). Indeed, it is sufficient to put

$$\psi(x) = -\frac{d}{dx} \ln \gamma(x) = -\frac{\gamma'(x)}{\gamma(x)}, \qquad x \in R_1.$$

On the contrary, if one starts from the M–estimator defined by some ρ–function, then the corresponding P–estimator is given by

$$\gamma(x) = \exp\{-\rho(x)\}, \qquad x \in R_1.$$

6i. W–estimators of the location parameter.

Let $X_1, \dots, X_n$ be a random sample from the distribution $F(x - \beta)$, where $x \in R_1$ and $\beta \in R_1$ is an unknown location parameter. Then the *W-estimator* of β is defined as that value $\widehat{W}_n$, which satisfies the equation

$$\widehat{W}_n = \frac{\sum_{i=1}^{n} X_i \, w\left(X_i - \widehat{W}_n\right)}{\sum_{i=1}^{n} w\left(X_i - \widehat{W}_n\right)}, \tag{47}$$

where $w : R_1 \mapsto R_1$ is a weight function. It seems that this type of robust estimators has been at the first time studied by Tukey (1977). If in the model the scale appears as a nuisance parameter, then $W-$ estimator $\widehat{W}_n$ should satisfy the equation

$$\widehat{W}_n = \frac{\sum_{i=1}^{n} X_i \, w\left(\frac{X_i - \widehat{W}_n}{\widehat{S}_n}\right)}{\sum_{i=1}^{n} w\left(\frac{X_i - \widehat{W}_n}{\widehat{S}_n}\right)}, \tag{48}$$

where $\widehat{S}_n$ is a (proper) estimator of the scale parameter. The problem whether to solve (48) simultaneously for both parameters has been briefly discussed earlier for $M-$ estimators of location and the same remark is valid here.

It follows immediately from (47) that one can obtain $\widehat{W}_n$ also as the solution of the equation

$$\frac{\sum_{i=1}^{n} (X_i - W_n) \, w\,(X_i - W_n)}{\sum_{i=1}^{n} w\,(X_i - W_n)} = 0. \tag{49}$$

But this is nothing else than the equation for the $M-$ estimator defined by the $\psi-$ function of the form $\psi(x) = x\,w(x)$, $x \in R_1$. Therefore, most of the properties of $W-$ estimators can be established using the theory derived for $M-$ estimators when taking the special form of $\psi(x)$ into account.

It is interesting to dwell for a moment on the question of computing of $W-$ estimators, especially because it is one of the basic ways how to compute also $M-$ estimators. Let $\widehat{W}_n^{(0)}$ be any initial estimator of β

and $\widehat{S}_n^{(0)}$ any initial estimator of the unknown scale parameter. Let us form the following recursive scheme, i.e.,

$$\widehat{W}_n^{(j+1)} = \frac{\sum_{i=1}^{n} X_i \, w\left(\frac{X_i - \widehat{W}_n^{(j)}}{\widehat{S}_n^{(0)}}\right)}{\sum_{i=1}^{n} w\left(\frac{X_i - \widehat{W}_n^{(j)}}{\widehat{S}_n^{(0)}}\right)}, \qquad j = 0, 1, \ldots . \tag{50}$$

We can continue in the iterations until, e.g., two consecutive values of the estimator do not differ too much, i.e., more precisely, until $\left|\widehat{W}_n^{(j+1)} - \widehat{W}_n^{(j)}\right| \leq \tau$, where $\tau > 0$ is some tolerance constant fixed in advance. The main reason why we discuss this approach in details is that this is the way how to compute $M-$estimators. In fact, it is sufficient to notice that if we use the weight function $w(x)$ of the form $w(x) = \psi(x)/x$, then we will obtain the desired solution. To be more precise, let us point out that the procedure just described is a special case of so called iteratively reweighted least squares method, which will be described in more details in the paragraph devoted to the computation of the $M-$estimators for linear model.

6j. Minimum distance estimators.

Let $X_1, \ldots, X_n$ be a random sample from the distribution $F_\theta(x)$, $\theta \in \Theta \subseteq R_1$, $x \in R_1$, and let $F_n(x)$ be the corresponding empirical distribution function. Let Π denotes some measure of discrepancy (badness of fit, distance between two probability distributions etc.), e.g., Kolmogorov – Smirnov statistic, Cramer – von Mises statistic, Hellinger distance etc. Then the *minimum distance estimator*, called sometimes also *D-estimator*, is defined as that value $\widehat{\theta}^{MD}$, for which $\Pi\left(F_n, F_\theta\right)$ is minimal, i.e., for which

$$\Pi\left(F_n, F_{\widehat{\theta}^{MD}}\right) = \inf_{\theta \in \Theta} \Pi\left(F_n, F_\theta\right). \tag{51}$$

Despite the fact that these estimators are not frequently used, results from both theoretical and simulation studies seem to be very promising. For more details and conditions necessary for proving the consistency, asymptotic normality etc. see Vajda (1984) or (1988).

7. LINEAR REGRESSION

7a. Introduction.

In this paragraph we will consider a linear regression model

$$\underset{\sim}{Y} = \underset{\sim}{X}\,\underset{\sim}{\beta}^0 + \underset{\sim}{e},$$

where $\forall n \in N$, $p \in N$, $\underset{\sim}{Y} = (Y_1, \ldots, Y_n)'$ is a response variable, $\underset{\sim}{X} = (x_{ij})_{i=1,\ldots,n}^{j=1,\ldots,p}$ is a design matrix, $\underset{\sim}{\beta}^0 = (\beta_1^0, \beta_2^0, \ldots, \beta_p^0)'$ is the "true" value of the regression coefficients and $\underset{\sim}{e} = (e_1, \ldots, e_n)'$ is a vector formed by an initial part of a sequence $\{e_i\}_{i=1}^{\infty}$ of i.i.d. r.v.'s distributed according to a distribution function F. Some authors instead of speaking about response variable and design matrix etc., use the names dependent variable or regresant for $\underset{\sim}{Y}$, see Sen and Srivastava (1990), e.g., and matrix of regressors, factors (then they also speak about the factor space), predictor variables, carriers or independent variables for $\underset{\sim}{X}$.

To introduce the idea of robustness we will assume that $F \in \mathcal{F}$ and the results will be derived with respect to $\mathcal{F}$. But the results are sometimes achieved under a compromise approach, let us say a mixed approach. For finite samples we do not specify F, at least "not too much", and we look for such properties as high breakdown point, small gross-error sensitivity, small local-shift sensitivity etc. On the other hand asymptotic properties, especially consistency and normality, are sometimes proved under rather narrow conditions as symmetry and unimodality of F etc. However, thinking about it twice one has to admit that there is a grain of logic in such an approach. While for small samples we are not able to determine F very precisely, large samples may reveal all misguided assumptions about the shape of F like a microscope.

It may be said that there are mainly two different (mathematical) approaches to the regression models. In the first one the design matrix $\underset{\sim}{X}$ is fixed, non-random, i.e., fixed matrix of constants. In the second one, on the contrary, $\underset{\sim}{X}$ is considered to be random and the distribution function F is assumed to be d.f. on R_{p+1}, $F(\underset{\sim}{t}) = P(x_{11} < t_1, x_{12} < t_2, \ldots, x_{1p} < t_p, e_1 < t_{p+1})$. This means that the design matrix plus vector of errors $(\underset{\sim}{X}, \underset{\sim}{e})$ are interpreted as the realizations of n independent, identically distributed $(p+1)$-dimensional random vectors $\{(\underset{\sim}{x}', \underset{\sim}{e})', i = 1, \ldots, n\}$, forming the rows of $\underset{\sim}{X}$.

It is a question rather for a philosophical discussions how this difference is relevant for practical purposes, i.e., for an application of results for finite sample situation. However, the assumptions in one approach usually correspond (rather closely) with the assumptions in the second approach for similar results. To give examples let us remind that one very frequent assumption in the "nonrandom" approach is that $\sup_{n \in N} \| \underset{\sim}{x}'_n \| < \infty$, sometimes weakened to existence of

$$\lim_{n \to \infty} \frac{1}{n} \sum_{i=1}^{n} \underset{\sim}{x}_i .$$

As a corresponding assumption in "random" approach we may view a requirement of $\mathsf{E}_F \| \underset{\sim}{x} \| < \infty$. Similarly, we have the correspondence

$$\lim_{n \to \infty} \frac{1}{n} \underset{\sim}{X}' \underset{\sim}{X} = \underset{\sim}{Q}, \qquad \underset{\sim}{Q} \text{ regular,}$$

in the "non-random" case and

$$\lim_{n \to \infty} \frac{1}{n} \mathsf{E}_F (\underset{\sim}{x} \underset{\sim}{x}') = \underset{\sim}{Q}, \qquad \underset{\sim}{Q} \text{ regular,}$$

in the "random" case.

For a function $\rho: R_1 \to R_1$ we will define an M-estimator by

$$\hat{\underset{\sim}{\beta}}^{\rho} = \arg \min_{\underset{\sim}{\beta} \in R_p} \sum_{i=1}^{n} \rho \left(\sigma^{-1} (Y_i - \underset{\sim}{x}'_i \underset{\sim}{\beta}) \right) . \tag{52}$$

To be in a full analogy with the location submodel, we should assume that $\rho: R_{p+1} \to R_1$, $Y_i \in R_1$, $\underset{\sim}{\beta} \in R_p$, but it would be a somewhat superfluous because we always assume dependence of ρ on $\underset{\sim}{\beta}$ through the residuals only. On the other hand, for the random approach some authors do assume $\rho: R_{p+1} \to R_1$, $\underset{\sim}{x}_i \in R_p$, $Y_i - \underset{\sim}{x}'_i \underset{\sim}{\beta} \in R_1$, see Maronna and Yohai (1981) or Hampel et. al. (1986). For $\rho(t) = t^2$ we have $\hat{\underset{\sim}{\beta}}^{\rho} = \hat{\underset{\sim}{\beta}}^{LS}$. Let us recall that $\hat{\underset{\sim}{\beta}}^{LS}$ is the maximum likelihood estimator when F is the normal distribution; hence for a general ρ, the M-estimator is similar to the maximum likelihood estimator. Assuming ρ differentiable and denoting $\psi(t) = d\rho(t)/dt$, we may look for $\hat{\underset{\sim}{\beta}}^{\rho}$ by solving for β set of equations

$$\sum_{i=1}^{n} \psi \left(\sigma^{-1} (Y_i - \underset{\sim}{x}'_i \underset{\sim}{\beta}) \right) x_{ik} = 0, \qquad k = 1, \ldots, p. \tag{53}$$

There exist at least three possibilities we can follow. The first one is to estimate (robustly) σ "separately" from estimation of β in a two stage procedure (see preceding discussion). Denoting the i-th residual $Y_i - x_i' \beta$ by $e_i(\beta)$ for $\beta \in R_p$, we can for an apriori evaluated robust and scale invariant estimator $\hat{\beta}^0$ compute $\left\{ e_i(\hat{\beta}^0) \right\}, i = 1, \ldots, n$ and then estimate σ for example by

$$\hat{\sigma}_{MAD} = 1.483 \operatorname*{med}_i \left\{ \left| e_i(\hat{\beta}^0) - \operatorname*{med}_j \left\{ e_j(\hat{\beta}^0) \right\} \right| \right\}.$$

For more details and some other estimator of scale see chapter 6f. But since $\hat{\sigma}_{MAD}$ is unbiased for the normal distribution and high robust in the case of heavy-tailed distributions, the use of $\hat{\sigma}_{MAD}$ is preferable at least at the beginning of the iterative process. Its lack of efficiency is repaid by a safe behavior in the case of contaminated data, which is particularly useful when starting the analysis. The estimator $\hat{\sigma}_{MAD}$ is then substituted into (53) instead of σ and $\hat{\beta}^\rho$ can be found.

We may add to (53) another equation for σ. This is frequently performed in such a way that (52) is modified to

$$\left(\hat{\beta}^{\rho'}, \hat{\sigma}^\rho \right)' = \operatorname*{arg\,min}_{(\beta', \sigma)' \in R_{p+1}} \sum_{i=1}^n \rho \left(\sigma^{-1} (Y_i - x_i' \beta) \right) \sigma + a\sigma.$$

Then instead of (53) one must solve following system of equations, i. e.,

$$\sum_{i=1}^n \psi \left(\sigma^{-1} (Y_i - x_i' \beta) \right) x_{ik} = 0, \qquad k = 1, \ldots, p, \tag{54}$$

and

$$\sum_{i=1}^n \chi \left(\sigma^{-1} (Y_i - x_i' \beta) \right) = a, \tag{55}$$

where

$$\chi(t) = t\psi(t) - \rho(t) \quad \text{and} \quad a = \int \chi(x) \, d\,\Phi(x). \tag{56}$$

Sometimes only equations (54) and (55) are used to determine $(\hat{\beta}^{\rho'}, \hat{\sigma}^\rho)'$ without relation (56), i. e., one then imposes some (reasonable) assumptions on the shape of $\rho\,(\psi)$ and χ separately.

Since most of the properties of M– estimators were explained for the location/scale problem, let us focus only on the special features of them concerning the regression problem. One may notice that equation (54) and the analogous equation for estimation of the location parameter (31), see also (35), differ by the term x_{ik}. This is a reason why a bad behaviour of the estimator $\hat{\underset{\sim}{\beta}}^{\rho}$ may be caused not only by *outliers*, i. e., the points which are deviated in the response variable, but also by points which are (considerably) far away from the bulk of data in the factor space. Such points are usually denoted (called) *leverage points*. It is clear that they carry a lot of information. Indeed, when they belong to the model, they increase considerably its reliability and may be called "golden points". On the contrary, when they are out of the model they give a hint of the character and the amount of the contamination. But, a fatal error may happen when one uses a method which interprets the leverage points of the second type wrongly as those of the first one. In the opposite case, i. e., when we consider erroneously some leverage points of the first type to be of the second one, and as is usual we reject (delete) them from the data set or weight them down, we loose "only" a part of the useful information, usually without fatal consequences. We have used the word "usually" in the previous sentence to say that there are situations, and not too complicated ones, when rejection of good leverage points may lead to completely wrong results. That is why we should use – as it will be mentioned later – diagnostic tools for searching for such influential points.

Due to the above discussed role of the factors x_{ik} in (54) it became clear rather soon that the ψ–function of the Huber type cannot cope with (wrong) leverage points. However, for some time statisticians believed that redescending ψ functions may cope with the problem better. Later on, when it turned out that this was a false hope, suggestions appeared "to weight the position of observations in the factor space" and the set of equations (54) was replaced by

$$(57) \qquad \sum_{i=1}^{n} \psi\left(\sigma^{-1}\left(Y_i - \underset{\sim}{x}_i' \underset{\sim}{\beta}\right)\right) x_{ik} w(\underset{\sim}{x}_i) = 0, \qquad k = 1, \ldots, p.$$

Well known and still used weights were proposed by Merill and Schweppe, Mallows, Hampel and Krasker, Krasker and Welch and many others, see Hampel et al. for detailed discussion, e. g. However, Maronna (1976) proved an assertion which says: "If the M– estimator is given as the solution of (57) and (55), there exists at least one solution which has breakdown point as low as $(p+1)^{-1}$. Let us recall that p denotes dimensionality of the model.

This result has been an inspiration for the rather natural question, i.e., how to establish the break-down point of the regression coefficients estimator. This point is discussed in paragraph 7f.

It is interesting to realize that the classical statistics offers for estimation of the parameters in the regression model only a few methods, most typical being the LSE and LAD (least absolute deviation estimator) ones. On the other hand, after introduction of the M-estimators and some other classes of robust estimators, which will be discussed later on, we will have at our disposal a few whole families of methods given by the families of ρ (ψ) functions, i.e., Andrews', Hampel's, Huber's, Tukey's etc. This brings us on the other hand one rather a big problem, i.e., when we will start to process the real data, we will be faced with the problem of selection of the most appropriate ρ (ψ) function. We will briefly return to this topic later on in paragraph 7k.

7b. Invariance.

Although the present explanation of regression model estimation cannot pretend to be (mathematically) exhaustive, let us mention at least some important theoretical notions connected with. One of them is the *invariance* of the estimators. To simplify the notation we will denote for a while by $\hat{\beta}(\underset{\sim}{Y}, \underset{\sim}{X})$ an estimator obtained by a given method from the data which consists of response variable $\underset{\sim}{Y}$ and design matrix $\underset{\sim}{X}$.

When discussing regression coefficient estimators, we speak usually about the following three types of invariance:

i) An estimator $\underset{\sim}{\hat{\beta}}$ is called ***regression equivariant*** if for any $\underset{\sim}{v} \in R_p$

$$\hat{\beta}(\underset{\sim}{Y} + \underset{\sim}{X}\,\underset{\sim}{v}, \underset{\sim}{X}) = \hat{\beta}(\underset{\sim}{Y}, \underset{\sim}{X}) + \underset{\sim}{v}.$$

ii) An estimator $\underset{\sim}{\hat{\beta}}$ is said to be ***scale invariant*** if for any $c \in R_1$

$$\hat{\beta}(c\underset{\sim}{Y}, \underset{\sim}{X}) = c\hat{\beta}(\underset{\sim}{Y}, \underset{\sim}{X}).$$

iii) We say that an estimator $\underset{\sim}{\hat{\beta}}$ is ***affine equivariant*** if for any $p \times p$ nonsingular matrix $\underset{\sim}{A}$

$$\hat{\beta}(\underset{\sim}{Y}, \underset{\sim}{X}\,\underset{\sim}{A}) = \underset{\sim}{A}^{-1}\hat{\beta}(\underset{\sim}{Y}, \underset{\sim}{X}).$$

Sometimes the last property is required to hold for orthogonal matrices only.

Remarks:

i) It is easy to show that if the method depends on the data only through the residuals, then it is regression equivariant, scale invariant and affine equivariant. To verify that, it is sufficient to realize that the residuals corresponding to the data $(\underset{\sim}{Y}, \underset{\sim}{X})$ and any $\beta_1 \in R_1$ are the same as the residuals corresponding to the data $(\underset{\sim}{Y} + \underset{\sim}{X}\underset{\sim}{v}, \underset{\sim}{X})$ and $\beta_1 + \underset{\sim}{v}$, i.e.,

$$\hat{\underset{\sim}{\beta}}(\underset{\sim}{Y} + \underset{\sim}{X}\,\underset{\sim}{v}, \underset{\sim}{X}) = \hat{\underset{\sim}{\beta}}(\underset{\sim}{Y}, \underset{\sim}{X}) + \underset{\sim}{v}.$$

Similar lines lead to the verification that any method which depends solely on the residuals is also scale invariant and affine equivariant.

ii) Analogously may be shown, due to the fact that regression equivariance and scale invariance "do not change" the design matrix, that any method which depends only on the residuals and on the elements of the design matrix is invariant in these two senses.

iii) On the other hand, it can be easily found that M-estimators which weight down positions in the factor space will be affine equivariant if and only if $w(\underset{\sim}{x}\,\underset{\sim}{A}) = w(\underset{\sim}{x})$ for any regular matrix $\underset{\sim}{A}$. But this may hold only either for very strange, and probably useless, weight functions or for a case when we restrict our notion of affine equivariance to the orthonormal matrices and w depends on $\|\underset{\sim}{x}\|$.

7c. Consistency.

Another important concept, maybe even more important than the invariance, is *consistency*. There exist only very special situations when we put up, or even prefer, the estimators which are not consistent. That is why we try to find necessary and sufficient conditions for any new estimator to be, either weakly or strongly, consistent. In the case when we are able to give only sufficient conditions we permanently try to find some weaker ones. This endeavour may lead sometimes to a strange situation; the problem will be discussed later. What is known now is that for all reasonable ψ-functions which are usually applied, corresponding M-estimator is, at least weakly, consistent. Presumably the most frequently cited paper is that of Maronna and Yohai (1981). We will quote their result briefly. However, let us change at first our notation for a while. Maronna and Yohai have included the weight function, remember (57), into the ψ function and assumed $\psi : R_p \times R_1 \to R_1$ and then they defined the M-estimator of the regression model coefficients as a solution of

$$\sum_{i=1}^{n} \psi\left(\underset{\sim}{x}_i, \sigma^{-1}(Y_i - \underset{\sim}{x}_i' \underset{\sim}{\beta})\right) x_{ik} = 0, \qquad k = 1, \ldots, p, \tag{58}$$

and

$$\sum_{i=1}^{n} \chi\left(\sigma^{-1}(|Y_i - \underset{\sim}{x}_i' \underset{\sim}{\beta}|)\right) = 0, \tag{58'}$$

where $\chi : R_1 \to R_1$ is given by (56). Moreover, let us assume that the design matrix is a realization of n p-dimensional r. v.'s and the distribution function F will for a moment be understood as a joint distribution function of any element of the sequence of i. i. d. r. v.'s $\{(\underset{\sim}{x}_i', e_i)'\}_{i=1}^{\infty}$.

Let us assume:

i) $\mathsf{E}_F \|\underset{\sim}{x}\| \sup_{u \in R_1} |\psi(\underset{\sim}{x}, u)| < \infty$;

ii) $\sup\left\{P(\underset{\sim}{x}_1' \underset{\sim}{\beta} = 0) : \underset{\sim}{\beta} \neq 0\right\} < \chi(u_0)/(a + \chi(u_0))$ where $u_0 \in R_1$ is defined so that $u_0 > 0$ and $\psi(\underset{\sim}{x}', u)/u > 0$ for any $\underset{\sim}{x} \in R_p$;

iii) $\sup\left\{P(\alpha y + \underset{\sim}{x}' \underset{\sim}{\beta} = 0) : \alpha \in R_1,\ \underset{\sim}{\beta} \in R_p,\ |\alpha| + \|\underset{\sim}{\beta}\| \neq 0\right\} < \frac{b}{a+b}$, where $a = -\chi(0)$ and $b = \chi(+\infty)$. Moreover, let $\left((\underset{\sim}{\beta}^0)', \sigma^0\right)'$ be solution of

$$\mathsf{E}_F\, \psi\left(\underset{\sim}{x}, \sigma^{-1}(\underset{\sim}{Y} - \underset{\sim}{X}\underset{\sim}{\beta})\right) \underset{\sim}{x} = 0$$

and

$$\mathsf{E}_F\, \chi\left(\underset{\sim}{x}, \sigma^{-1}(|\underset{\sim}{Y} - \underset{\sim}{X}\underset{\sim}{\beta}|)\right) \underset{\sim}{x} = 0.$$

Finally let $\left\{\left(\hat{\underset{\sim}{\beta}}_n', \hat{\sigma}_n\right)'\right\}_{n=p}^{\infty}$ be solutions of (58) and (58').

Then $\left(\hat{\underset{\sim}{\beta}}_n', \hat{\sigma}_n\right)'$ is strongly consistent, i. e., almost surely converging to $\left((\underset{\sim}{\beta}^0)', \sigma^0\right)'$. Please realize that in this paragraph $\underset{\sim}{x}$ is an element of R_p and not the matrix $\underset{\sim}{X}$.

7d. Asymptotic normality.

The step following the proof of the consistency is usually an attempt to prove asymptotic normality. One very elegant way how to proceed is to find an asymptotic representation for $\sqrt{n}\left(\hat{\beta}^{\rho} - \beta^{0}\right)$. One of the first results of this type can be found in Huber (1967). An example of "the present state of art", i. e., how far the theory has proceeded during 20 years, is a result which is due to Jurečková (1988). It reads as follows. Let $\psi : R_1 \to R_1$, $\psi = \psi_1 + \psi_2 + \psi_3$, fulfill following the conditions:

i) ψ_1 is a step-function with a finite number of jump-points;
ii) ψ_2 is absolutely continuous and ψ_2' is a step-function with a finite number of jump-points;
iii) ψ_3 is absolutely continuous with an absolutely continuous derivative ψ_3' and there exist $\nu > 1$ and $\delta > 0$ such that

$$\int_{R_1} |\psi_3'(x+u)|^{\nu}\, dF(x) < K < \infty \qquad \text{for } |u| < \delta,$$

and put

$$\gamma = \gamma_1 + \gamma_2 = \int_{R_1} f(x)\, d\psi_1(x) + \int_{R_1} (\psi_2' + \psi_3')\, dF(x),$$

where $f(x) = F'(x)$. We assume that $f(x)$ exists at least in a neighborhood of jump-points of ψ_1. Moreover, let

$$\lim_{n\to\infty} \frac{1}{n} X' X = Q, \qquad Q \quad \text{regular.}$$

Further, let $\mathsf{E}\, e_1^2 = \sigma^2$, $0 < \sigma^2 < \infty$. and let us denote

$$\Psi\left(\frac{e}{\sigma}\right) = \left(\psi\left(\frac{e_1}{\sigma}\right), \psi\left(\frac{e_2}{\sigma}\right), \ldots, \psi\left(\frac{e_n}{\sigma}\right)\right)'.$$

Then

$$\sqrt{n}\left(\hat{\beta}^{\rho} - \beta^{0}\right) = \frac{1}{\gamma\sqrt{n}}\, Q^{-1} X' \Psi\left(\frac{e}{\sigma}\right)\sigma + o_p(1), \qquad \text{as} \quad n \to \infty.$$

Let us note that the main tool of this approach to prove asymptotic normality is a strong approximation of some empirical process in the sense of Csörgö and Révész' (1981) book and asymptotic linearity of this empirical process.

Asymptotic normality of $\sqrt{n}\left(\left(\hat{\beta}'_n, \hat{\sigma}_n\right)' - \left(\beta^{0\prime}, \sigma^0\right)'\right)$ for the "random" approach may be found in Maronna and Yohai (1981). Although their proof seems to be "easier" than in the "non-random" case, it is far less transparent. Notice also, that their assumptions in the framework of the "random" model for the proof of consistency and asymptotic normality are of such a type that it is not very easy to find a heuristic interpretation for them.

7e. Recursive evaluation.

Sometimes we are in the situation that we get one observation after another and we would like to recalculate the estimator $\hat{\beta}^{\rho,n-1}$ obtained for $n-1$ observations to a new estimator $\hat{\beta}^{\rho,n}$ for n observation. It is indeed possible to do this for any M-estimator which is generated by a ψ-function which is (at least approximately) linear around the origin. Using Taylor's expansion for the equation determining the M-estimator and neglecting elements of higher orders we can derive:

$$\hat{\beta}^{\rho,n} \cong \hat{\beta}^{\rho,n-1} - H^{-1}(n)\nabla_{\beta} C_{\rho}(n, \beta)_{\beta=\hat{\beta}^{\rho,n-1}}, \tag{59}$$

where

$$C_{\rho}(n, \beta) = \sum_{i=1}^{n} \rho\left(\sigma_0^{-1}(Y_i - x_i'\beta)\right)\sigma_0^2,$$

$$\nabla_{\beta} C_{\rho} = \frac{\partial C_{\rho}}{\partial \beta}, \quad \text{and} \quad H(n) = \nabla_{\beta}\nabla_{\beta} C_{\rho}(n, \beta)_{\beta=\hat{\beta}^{\rho,n-1}}$$

Expanding $\nabla_{\beta} C_{\rho}(n, \beta)_{\beta=\hat{\beta}^{\rho,n-1}}$ into a Taylor's serie and using the recursive representation for $H(n)$, which again can be found relatively easily via the Taylor expansion and well known results from the linear algebra, of the form

$$H^{-1}(n) = H^{-1}(n-1) - \rho'\left(\sigma_0^{-1}\left(Y_n - x_n'\hat{\beta}^{\rho,n-1}\right)\right) \cdot$$
$$\cdot \frac{H^{-1}(n-1)X_n X_n' H^{-1}(n-1)}{1 + x_n' H^{-1}(n-1) x_n \rho'\left(\sigma_0^{-1}(Y_n - x_n'\hat{\beta}^{\rho,n-1})\right)},$$

we may rewrite (59) in a more convenient form, i.e.,

$$\hat{\underset{\sim}{\beta}}^{\rho,n} \cong \hat{\underset{\sim}{\beta}}^{\rho,n-1} + \underset{\sim}{H}^{-1}(n-1)\underset{\sim}{x}_n\psi\left(\sigma_0^{-1}\left(Y_n - \underset{\sim}{x}_n'\hat{\underset{\sim}{\beta}}^{\rho,n-1}\right)\right)\sigma_0.$$

The last relation holds precisely for LSE, but this not the case of the M-estimators in general. Nevertheless, practical experience shows that this approximation is precise enough for the M-estimator which are generated by a ψ-function which is, at least approximately, linear around the origin. Therefore, for practical application of this approach it is necessary to be aware that the relation is only an approximation and hence from time to time we had to evaluate precise solution of (53) instead of forming (59).

7f. Selection of the ρ-function.

Before turning our attention to the other types of estimators, let us concentrate on the problem mentioned above, i.e., the choice of the most appropriate ρ-function in the practice. Since conditions for the consistency of (M-) estimators are rather weak, we may use the whole spectrum of methods for the same data to estimate a regression model. Then it may happen that we obtain rather different estimates for the model and quite naturally the question "Which one of them should be selected?" arises. Two of the simplest possible answers are based on the following ideas.

At first we have assumed that the estimated model is "close" to the "true" one. The "true" model holds (or should hold) for all data, and hence it should be valid also for any subsample of the data. An estimator of model for any (or at least for some reasonably – naturally not too much – restricted) subsamples should then also be "close" to the "true" model and hence also "close" to the originally estimated model (for the whole data). This means, in other words, that the maximal (studentized) difference between the model for the whole data set and models for any subsample (let us say of size $n-1$) has to be statistically insignificant.

Under not very restrictive conditions, which cover all usually used families of ρ-functions generating M-estimators, it is possible to show that for $n \to \infty$

$$n\left\{\hat{\underset{\sim}{\beta}}^{\rho}\left(\underset{\sim}{Y}^{(q)}, \underset{\sim}{X}^{(q)}\right) - \hat{\underset{\sim}{\beta}}^{\rho}(\underset{\sim}{Y}, \underset{\sim}{X})\right\} =$$

$$= -\,\mathsf{E}_F^{-1}\psi'(e_1)\,\underset{\sim}{Q}^{-1}\,\underset{\sim}{x}_q'\psi(r_q) + o_p(1),$$

where $\hat{\beta}^{\rho}\left(Y^{(q)}, X^{(q)}\right)$ is the M-estimator (determined by ρ) for data from which q-th observation has been deleted, $Q = \lim_{n\to\infty} n^{-1} X'X$, Q regular, x_q' is the q-th row of the design matrix X and $r_q = r_q(\hat{\beta}^{\rho}) = Y_q - x_q'\hat{\beta}^{\rho}(Y, X)$, i.e., the q-th residual in the model for the whole data. Using Šidák's (1967) inequality, one may derive a corresponding critical region for the test of significance of studentized change of norm of the estimator. We do not have time and space to go into details, hence let us only refer to Víšek (1991 a, b).

Secondly, it may happen that differences of models which were established using different ρ-functions seem to be "optically" large but they are not statistically significant. Please, notice that it may also be vice versa. Using the asymptotic representation described above for $\sqrt{n}\left(\hat{\beta}^{\rho} - \beta^{0}\right)$, one may derive an asymptotic representation for difference of two models in the form

$$(60)\quad \frac{1}{\sqrt{n}}\sum_{i=1}^{n}\left(r_i^{(1)} - r_i^{(2)}\right) =$$

$$= \frac{\sigma}{\sqrt{n}}\sum_{i=1}^{n}\left\{\frac{1}{n}\sum_{j=1}^{n}\sum_{r=1}^{p}\sum_{s=1}^{p} x_{jr}\, q_{rs}\, x_{is}\right\}\left[\frac{\gamma_{(1)}}{\gamma_{(2)}}\frac{\psi_{(2)}\left(\frac{e_i}{\sigma}\right)}{\psi_{(1)}\left(\frac{e_i}{\sigma}\right)}\right] + o_p(1),$$

as $n \to \infty$, where $q_{rs} = \left\{Q^{-1}\right\}_{rs}$, $r_i^{(k)}$, $k = 1, 2$, is the i-th residual in the model estimated using $\psi^{(k)}$ and $\gamma_{(k)} = \mathsf{E}_F \psi_{(k)}'(t)$. Then making use of a small-sample asymptotic result one may find a critical region for (60). For details see Víšek (1991 c).

A quite different solution of the problem may consist in taking an average, or Bayes estimator, from the whole family of estimators. Study of such an approach started only recently and the first experiences show that this approach is promising, but computationally very intensive (see Víšek (1991 d) or He et al. (1990)).

7g. The estimators with high breakdown point.

Let us turn back to the question of an attainable level of the break-down of the estimators. It seems that the first author who introduced a regression estimator with asymptotic break-down point equal to 50 % was Siegel (1982). However, his method of repeated medians was rather clumsy and has been never used in the practice. Nevertheless, new proposals followed soon. The first of them, and probably the most popular till now, is Rousseeuw's (1984) least median of squares (LMS), which is defined as

$$\hat{\underset{\sim}{\beta}}^{LMS} = \arg \min_{\underset{\sim}{\beta} \in R_p} \operatorname*{med}_i \left\{ \left[Y_i - \underset{\sim}{x}_i' \underset{\sim}{\beta} \right]^2 \right\}.$$

Its disadvantage is lack of efficiency which is caused by $n^{-1/3}$ rate of consistency. If the distribution function F is symmetric and strongly unimodal with density f, then the distribution of LMS estimator converges weakly to

$$\mathcal{L}\left(\sqrt[3]{n} \left(\hat{\underset{\sim}{\beta}}^{LMS} - \underset{\sim}{\beta}^0 \right) \right) \longrightarrow \underset{\sim}{\mathcal{L}}(C\tau),$$

where $C = \left(f(F^{-1}(0.75)) \, f'(F^{-1}(0.75))^2/2 \right)^{-1/3}$ and τ is the random time s for which $W(s)+s^2$ attains its minimum, $W(s)$ being a standard two-sided Brownian motion originating from zero. For proposals of the algorithms for evaluation of LMS see Joss and Marazzi (1991) or Tichavský (1991).

One way how to remove the disadvantage of lack of efficiency of LMS is the use of one-step M-estimator instead, preferably with redescending ψ. By a "redescending" ψ-function we understand such a function which "redescends" to zero, see for instance Hampel's ψ in the Table 1. Such an estimator may be defined as

$$\hat{\underset{\sim}{\beta}}^{OSM} = \hat{\underset{\sim}{\beta}}^{LMS} + (\underset{\sim}{X}'\underset{\sim}{X})^{-} \underset{\sim}{X}' \cdot$$

$$\cdot \left(\psi \left(\hat{\sigma}_{MAD}^{-1} e_1 \left(\hat{\underset{\sim}{\beta}}^{LMS} \right) \right), \ldots, \psi \left(\hat{\sigma}_{MAD}^{-1} e_n \left(\hat{\underset{\sim}{\beta}}^{LMS} \right) \right) \right)' \frac{\hat{\sigma}_{MAD}}{B(\psi, \Phi)}$$

with $B(\psi, \Phi) = \int \psi'(t) \, d\Phi(t)$ and Φ being the distribution function of the standard normal distribution. It is interesting that $\hat{\underset{\sim}{\beta}}^{OSM}$ inherits robust features of $\hat{\underset{\sim}{\beta}}^{LSM}$, but its asymptotic behavior is better than that of $\dot{\beta}^{\rho}$. Then we have, for a "reasonable" ψ,

$$\mathcal{L}\left(\sqrt{n} \left(\hat{\underset{\sim}{\beta}}^{OSM} - \underset{\sim}{\beta}^0 \right) \right) \longrightarrow \mathcal{N}\left(\underset{\sim}{0}, \underset{\sim}{V}(\psi, F) \underset{\sim}{Q}^{-1} \right),$$

where $\underset{\sim}{Q} = \lim_{n \to \infty} n^{-1} \underset{\sim}{X}' \underset{\sim}{X}$ is a positive definite $p \times p$ covariance matrix and

$$\underset{\sim}{V}(\psi, F) = \frac{\mathsf{E}_F \, \psi \left(\dfrac{t}{\hat{\sigma}_{MAD}} \right)}{\left[\mathsf{E}_F \, \psi' \left(\dfrac{t}{\hat{\sigma}_{MAD}} \right) \right]^2}.$$

Another possibility to improve the behavior of $\hat{\underset{\sim}{\beta}}^{LMS}$ directly has been proposed by Rousseeuw (1983). This estimator is known as the least trimmed squares estimator (LTSE) and is defined as

$$\hat{\underset{\sim}{\beta}}^{LTS} = \arg\min_{\underset{\sim}{\beta} \in R_p} \sum_{i=1}^{h} e_i^2(\underset{\sim}{\beta}),$$

where $e_i(\underset{\sim}{\beta})$ is the $i-$th order statistic among $e_i(\underset{\sim}{\beta}) = Y_i - \underset{\sim}{x}_i'\underset{\sim}{\beta}$, $i = 1, \ldots, n$, $h = [n/2] + [(p+1)/2]$ and $[a]$ denotes the largest integer which is less or equal to a. The $\hat{\underset{\sim}{\beta}}^{LTS}$ also has asymptotic breakdown point equal to 50 %, but under assumption of symmetry of F we can show that

$$\mathcal{L}\left(\sqrt{n}\left(\hat{\underset{\sim}{\beta}}^{LTS} - \underset{\sim}{\beta}^0\right)\right) \longrightarrow \mathcal{N}\left(\underset{\sim}{0}, \underset{\sim}{V}(LTS, F)\underset{\sim}{Q}^{-1}\right),$$

where

$$\underset{\sim}{V}(LTS, F) = \frac{8\int_0^q y^2\, dF(y)}{\left[1 - 4F^{-1}(0.75).f(F^{-1}(0.75))\right]^2}.$$

Please, do not confuse $LTSE$ with the trimmed least squares estimator $TLSE$, which will be introduced in Section 7h devoted to the $L-$estimators.

7h. *S*−estimators.

Let us reconsider LS once again. What we do when looking for the LS solution of the regression model is to search for

$$\arg\min_{\underset{\sim}{\beta} \in R_p} \sum_{i=1}^{n} \left(Y_i - \underset{\sim}{x}_i'\underset{\sim}{\beta}\right)^2. \tag{61}$$

If we put $\rho_{LS}(t) = t^2$ and try to find such $\underset{\sim}{\beta} \in R_p$ which solves

$$\frac{1}{n}\sum_{i=1}^{n} \rho_{LS}\left(\sigma^{-1}(Y_i - \underset{\sim}{x}_i'\underset{\sim}{\beta})\right) = 1 \tag{62}$$

and for which σ is minimal among all possible values, we obtain just the LSE. Really, from (61) we obtain $n^{-1}\sum_{i=1}^{n}(Y_i - \underset{\sim}{x}_i'\underset{\sim}{\beta})^2 = \sigma^2$ and hence if we minimize σ under constraint (62) this is equivalent to solving (61). It may be an inspiration for definition of another family of estimators – the S–estimators, where S stays for the scale minimizing. For a general ρ function S–estimators are defined as such $\hat{\underset{\sim}{\beta}}^{S,\rho} \in R_p$, for which σ in

$$(63) \qquad \frac{1}{n} \sum_{i=1}^{n} \rho\left(\sigma^{-1}(Y_i - \underset{\sim}{x}_i' \underset{\sim}{\beta})\right) = b$$

is minimal, b being prescribed apriori. For the practical purposes is recommended to put $b = \int \rho(t)\, d\Phi(t)$, the main reason being to guarantee high efficiency of the $\hat{\underset{\sim}{\beta}}^{S,\rho}$ for normally distributed errors.

It is clear that we may interpret σ as a function of $\underset{\sim}{\beta}$ and then to define the S–estimator also as

$$(63') \qquad \hat{\underset{\sim}{\beta}}^{S,\rho} = \arg\min_{\underset{\sim}{\beta} \in R_p} \sigma(\underset{\sim}{\beta})$$

under constraint that

$$\sum_{i=1}^{n} \rho\left(\sigma^{-1}(\underset{\sim}{\beta})(Y_i - \underset{\sim}{x}_i' \underset{\sim}{\beta})\right) = n\, b.$$

When studying the properties of $\underset{\sim}{\beta}^{S,\rho}$, we usually impose on ρ the following conditions:

i) ρ is symmetric, continuously differentiable and $\rho(0) = 0$;
ii) there exists $c > 0$ such that ρ is strictly increasing on $[0, c]$ and constant on (c, ∞).

The condition (ii) may imply the existence of more solutions of (62) for one fixed $\underset{\sim}{\beta} \in R_p$. Therefore, the definition of S–estimator had to be modified as follows. Put for any fixed $\underset{\sim}{\beta} \in R_p$

$$\sigma(\underset{\sim}{\beta}) = \sup\left\{\sigma \mid \sigma \text{ solves } (62)\right\}.$$

Then

$$(63'') \qquad \hat{\underset{\sim}{\beta}}^{S,\rho} = \arg\min_{\underset{\sim}{\beta} \in R_p} \sigma(\underset{\sim}{\beta})$$

and

$$\hat{\sigma}^{S,\rho} = \sigma\left(\hat{\underset{\sim}{\beta}}^{S,\rho}\right).$$

Realize that then $\sigma(\underset{\sim}{\beta}) \geq \hat{\sigma}^{S,\rho}$ for any $\underset{\sim}{\beta} \in R_p$. Under an additional condition $\rho(c) = 2b$, which is due to the fact that $b = \mathsf{E}_\Phi \rho$, the corresponding $S-$estimator has asymptotically breakdown point 50 % and the precise value for finite sample size n is $([n/2] - p + 2)/n$. For this and following results concerning $S-$estimators see Rousseeuw and Leroy (1987), e. g.

Let us assume further that ρ is differentiable. Since for any $\underset{\sim}{\beta} \in R_p$ we have

$$\sum_{i=1}^{n} \rho\left(\sigma^{-1}(\underset{\sim}{\beta})(Y_i - \underset{\sim}{x}_i' \underset{\sim}{\beta})\right) = n\,b,$$

for $\sigma(\underset{\sim}{\beta})$, so that for $\hat{\sigma}$ we shall have

$$\sum_{i=1}^{n} \rho\left(\hat{\sigma}^{-1}(Y_i - \underset{\sim}{x}_i' \underset{\sim}{\beta})\right) \geq n\,b$$

for any $\underset{\sim}{\beta} \in R_p$ and at $\underset{\sim}{\hat{\beta}}^{S,\rho}$ it turns to the equality. So $\underset{\sim}{\hat{\beta}}^{S,\rho}$ minimizes

$$\sum_{i=1}^{n} \rho\left(\hat{\sigma}^{-1}(Y_i - \underset{\sim}{x}_i' \underset{\sim}{\beta})\right).$$

Taking partial derivatives with respect to β_k, $k = 1, \ldots, p$, we obtain

$$\sum_{i=1}^{n} \psi\left(\hat{\sigma}^{-1}(Y_i - \underset{\sim}{x}_i' \underset{\sim}{\hat{\beta}}^{S})\right) x_{ik} = 0, \tag{64}$$

and denoting $\rho(t) - n\,b$ by χ, we have also

$$\sum_{i=1}^{n} \chi\left(\hat{\sigma}^{-1}(Y_i - \underset{\sim}{x}_i' \underset{\sim}{\hat{\beta}}^{S})\right) = 0. \tag{65}$$

Please notice that it is necessary to interpret very carefully what we have derived. It is evident that the $S-$estimator is one of the possible solutions of (64) and (65), i. e., one of the possible values of the $M-$estimator given by ρ. But the above stated result about the breakdown point of $\underset{\sim}{\hat{\beta}}^{S,\rho}$ implies that the $S-$estimator is different from that solution of (64) and (65) which Maronna's result concerning breakdown point of $M-$estimators speaks about.

If in addition to the conditions (i) and (ii) $\psi(u)/u$ is nonincreasing for $u > 0$ and if $\mathsf{E}_F\|\underset{\sim}{x}\| < \infty$, of course in the case of random carriers, then

$$\left(\underset{\sim}{\hat\beta}^{S,\rho\prime}, \hat\sigma^{S}\right) \longrightarrow \left(\underset{\sim}{\beta}^{0\prime}, \sigma^0\right)' \qquad \text{a. s. as} \qquad n \longrightarrow \infty.$$

Analogously, if moreover $\int \psi' \, d\Phi > 0$, $\mathsf{E}_F \underset{\sim}{x}\,\underset{\sim}{x}'$ is nonsingular and $\mathsf{E}_F\|\underset{\sim}{x}\|^3 < \infty$, then

$$\mathcal{L}\left(\sqrt{n}\left(\underset{\sim}{\hat\beta}^{S,\rho} - \underset{\sim}{\beta}^0\right)\right) \longrightarrow \mathcal{N}\left(\underset{\sim}{0}, \mathsf{E}_F(\underset{\sim}{x}\,\underset{\sim}{x}')^{-1} \frac{\int \psi^2(t)\, d\Phi(t)}{\left[\int \psi'(t)\, d\Phi(t)\right]^2}\right)$$

and

$$\mathcal{L}(\sqrt{n}\,(\hat\sigma^{S,\rho} - \sigma^0)\,) \longrightarrow \mathcal{N}\left(\underset{\sim}{0}, \frac{\int (\rho(t) - b)\, d\Phi(t)}{\left[\int t\psi(t)\, d\Phi(t)\right]^2}\right).$$

Please, realize also here that as in the case when we have quoted Maronna – Yohai's result about consistency of $M-$estimators, that $x \in R_p$. If analogous assertion is given for the scheme with nonrandom carriers (design matrix) the assumption $\mathsf{E}\|\underset{\sim}{x}\| < \infty$ is to be modified to: $\lim_{n\to\infty} n^{-1}\sum_{i=1}^{n} x_{ik}$ exist and is finite for all $k = 1, \ldots, p$. Similarly, commonly used assumption for the nonrandom carriers scheme is that

$$\lim_{n\to\infty} \frac{1}{n} \underset{\sim}{X}'\underset{\sim}{X} = \underset{\sim}{Q}, \qquad \underset{\sim}{Q} \quad \text{regular},$$

what corresponds to the assumption that $\mathsf{E}_F \underset{\sim}{x}\,\underset{\sim}{x}'$, $\underset{\sim}{x} \in R_p$, is regular.

7i. $L-$estimators.

As we have seen earlier, the $M-$estimators of the regression coefficients are natural and straightforward generalizations of the $M-$estimators of location. Another very appealing family of location estimators is the family of $L-$estimators. This is especially due to the fact that on one side it is (relatively) easy to compute them and on the other hand they have properties (at least asymptotically) as good as the $M-$estimators. From the mid-sixties when Huber made a pioneer step introducing the $M-$estimators there were attempts to generalize the $L-$estimators to the regression schema, too. However, it took nearly fifteen years before two american econometricians Roger Koenker and Gilbert Bassett (1978) published in Econometrica the first really successful proposal which has attracted attention of many statisticians. Let us describe their idea in more details.

Let us denote for a fixed $\alpha \in (0,1)$

(66) $$\psi_\alpha(t) = \alpha - I_{\{t<0\}} \quad \text{and} \quad \rho_\alpha(t) = t\psi_\alpha(t),$$

where $I_{\{t<0\}}$ is the indicator function. Then

(67) $$\hat{\beta}_\alpha = \arg\min_{\beta \in R_p} \sum_{i=1}^{n} \rho_\alpha (Y_i - x_i' \beta)$$

will be called the α*-regression quantile* (one may notice that for the special case $X = \{x_{i1}\}_{i=1}^{\infty}$, $x_{i1} = 1 \ \forall i \in N$, we obtain the sample α-quantile - hence the name). The original proposal of Koenker and Bassett was to consider a linear combination (just as for location parameter) of regression quantiles. For the special case $\alpha = 1/2$ we obtain $\hat{\beta}_{1/2} = \hat{\beta}^{L_1}$ - LAD (least absolute deviations) estimator, sometimes called also L_1-norm estimator. Later on Ruppert and Carroll (1978, 1980) followed the idea of Koenker and Bassett and proposed a generalization of the α-trimmed mean. More precisely, they have advised to find for some fixed α_1, α_2, $0 < \alpha_1 < \alpha_2 < 1$

(68) $$\hat{\beta}^{TLS} = \arg\min_{\beta \in R_p} \sum_{i \in I_{\alpha_1,\alpha_2}} (Y_i - x_i' \beta)^2,$$

where $I_{\alpha_1,\alpha_2} = \left\{ i \middle| 1 \le i \le n,\ x_i'\hat{\beta}_{\alpha_1} < Y_i < x_i'\hat{\beta}_{\alpha_2} \right\}$.

At a first sight it seemed that computation of such kind of estimators would be rather difficult, or as it is today fashionable to say, computationally intensive. However, the authors of the proposal demonstrated that it is possible to use the well-known simplex method, which is easily available in software packages. Moreover, it turns out that $\hat{\beta}_\alpha$ as a function of α, i.e., $\hat{\beta}_\alpha : (0,1) \to R_p$, can attain only a finite number of values.

A convenient notation for employment of the simplex methods is the following one. Let us denote for any $\beta \in R_p$ $e_i(\beta) = Y_i - x_i' \beta$. Then (67) is equivalent to

$$\begin{aligned} \hat{\beta}_\alpha &= \arg\min_{\beta \in R_p} \left\{ \sum_{\{i:\, e_i(\beta) > 0\}} \alpha \cdot e_i(\beta) + \sum_{\{i:\, e_i(\beta) < 0\}} (\alpha - 1) \cdot e_i(\beta) \right\} \\ &= \arg\min_{\beta \in R_p} \sum_{i=1}^{n} w(e_i(\beta)) \cdot e_i(\beta), \end{aligned}$$

where $w(t) = \psi_\alpha(t) = \alpha - I_{\{t<0\}}$. Let us denote $\underset{\sim}{1} = (1, \ldots, 1)'$, and $\underset{\sim}{e}(\underset{\sim}{\beta}) = \left(e_1(\underset{\sim}{\beta}), \ldots, e_n(\underset{\sim}{\beta})\right)'$, $\underset{\sim}{e}^+(\underset{\sim}{\beta}) = \max\{\underset{\sim}{0}, \underset{\sim}{e}(\underset{\sim}{\beta})\}$, $\underset{\sim}{e}^-(\underset{\sim}{\beta}) = \max\left\{\underset{\sim}{0}, -\underset{\sim}{e}(\underset{\sim}{\beta})\right\}$, $\underset{\sim}{\beta}^+ = \max\{\underset{\sim}{0}, \underset{\sim}{\beta}\}$ and $\underset{\sim}{\beta}^- = \max\{\underset{\sim}{0}, -\underset{\sim}{\beta}\}$, where all maxima are taken coordinate wise. Then the previous equation may be rewritten in a form

$$\hat{\underset{\sim}{\beta}}_\alpha = \underset{\underset{\sim}{\beta} \in R_p}{\arg\min} \left\{\alpha \underset{\sim}{1}' \underset{\sim}{e}^+(\underset{\sim}{\beta}) + (1-\alpha)\underset{\sim}{1}' \underset{\sim}{e}^-(\underset{\sim}{\beta})\right\} \tag{69}$$

under the constraints

$$\underset{\sim}{e}(\underset{\sim}{\beta}) = \underset{\sim}{Y} - \underset{\sim}{X}\,\underset{\sim}{\beta}, \quad \underset{\sim}{e}^-(\underset{\sim}{\beta}) \geq 0, \quad \underset{\sim}{e}^+(\underset{\sim}{\beta}) \geq 0.$$

It has to be noticed that for the solution of the last minimization problem it is not necessary to use the classical simplex algorithm, which is rather slow. In fact, several much faster modification were proposed which take into account special structure of the problems, see Barrodale and Roberts (1973, 1974) or Koenker and d'Orey (1987) for details.

Also problem (68) may be easily modified so that one may use directly commonly available software. Before presenting this modification let us remark that we may "cut off" points also in an asymmetric way, i. e., we may calculate "upper" regression quantile for some α_1 $(0 < \alpha_1 \leq 0.5)$ and "lower" regression quantile for $\alpha_2 \neq 1 - \alpha_1$ $(0 < \alpha_1 < \alpha_2 < 1)$. If we denote by $\underset{\sim}{D}$ a (diagonal) matrix

$$\underset{\sim}{D} = \{d_{ij}\}_{i,j=1}^n, \quad d_{ij} = \delta_{ij} I_{\{\underset{\sim}{x}_i' \hat{\underset{\sim}{\beta}}_{\alpha_1} \leq Y_i \leq \underset{\sim}{x}_i' \hat{\underset{\sim}{\beta}}_{\alpha_2}\}}, \tag{70}$$

we may rewrite (68) as

$$\hat{\underset{\sim}{\beta}}^{TLS} = \underset{\underset{\sim}{\beta} \in R_p}{\arg\min} \sum_{i=1}^n \left(d_{ii}Y_i - d_{ii}\underset{\sim}{x}_i'\,\underset{\sim}{\beta}\right)^2,$$

so that $\hat{\underset{\sim}{\beta}}^{TLS}(\underset{\sim}{Y}, \underset{\sim}{X}) = \hat{\underset{\sim}{\beta}}^{LS}(\underset{\sim}{D}\underset{\sim}{Y}, \underset{\sim}{D}\underset{\sim}{X})$. In other words, $\hat{\underset{\sim}{\beta}}^{TLS}$ is a LS–estimator for the model

$$\underset{\sim}{D}\underset{\sim}{Y} = \underset{\sim}{D}\underset{\sim}{X}\,\underset{\sim}{\beta} + \underset{\sim}{D}\underset{\sim}{e},$$

i. e.,

$$\hat{\underset{\sim}{\beta}}^{TLS} = (\underset{\sim}{X}'\underset{\sim}{D}\underset{\sim}{X})^- \underset{\sim}{X}'\underset{\sim}{D}\underset{\sim}{Y},$$

where $(\underset{\sim}{X}'\underset{\sim}{D}\underset{\sim}{X})^-$ stays for a generalized inverse of $\underset{\sim}{X}'\underset{\sim}{D}\underset{\sim}{X}$. Please notice that the last formula is however only a theoretical relation which is not useful for practical computation.

7j. Asymptotics for L-estimators.

To further simplify the explanation, let us assume that the model includes the intercept, i.e., $x_{i1} = 1$ for $i = 1, \ldots, n$. Moreover, let d.f. F (of the error term) be absolutely continuous and symmetric, its density is positive and finite, f' exists and is bounded in a neighborhood of $F^{-1}(\alpha)$ and $F^{-1}(1-\alpha)$, $0 < \alpha < 1$. Moreover, let us assume for the design matrix that:

i) $\max_{1 \le i \le n,\, 2 \le j \le p} \frac{1}{\sqrt[4]{n}} |x_{ij}| = O(1)$ as $n \to \infty$ and

ii) $\lim_{n \to \infty} \frac{1}{n} \underset{\sim}{X}' \underset{\sim}{X} = \underset{\sim}{Q}$, $\quad Q$ is positively definite.

Under these condition $\hat{\underset{\sim}{\beta}}_\alpha$ is $\sqrt{n}$-consistent, more precisely

$$\sqrt{n}\left(\hat{\underset{\sim}{\beta}}_\alpha - \underset{\sim}{\beta}^0_\alpha\right) = O_p(1),$$

where $\underset{\sim}{\beta}^0_\alpha = \left(F^{-1}(\alpha) + \beta^0_1,\ \beta^0_2, \ldots, \beta^0_p\right)$. Moreover, the following asymptotic representation can be derived for $n \to \infty$, i.e.,

$$\sqrt{n}\left(\hat{\underset{\sim}{\beta}}_\alpha - \underset{\sim}{\beta}^0_\alpha\right) =$$

$$= \sqrt{n}\left[f\left(F^{-1}(\alpha)\right)\right]^{-1} \underset{\sim}{Q}^{-1} \sum_{i=1}^{n} d_{ii}\, \underset{\sim}{x}_i \psi_\alpha\left(e_i - F^{-1}(\alpha)\right) + O_p\left(\frac{1}{\sqrt[4]{n}}\right)$$

Let us recall that $\psi_\alpha(t)$ is given in (66).

Similarly, for $\hat{\underset{\sim}{\beta}}^{TLS}$ we have for $n \to \infty$ asymptotic representation of the form

$$\sqrt{n}\left(\hat{\underset{\sim}{\beta}}^{TLS}_{\alpha_1,\alpha_2} - \underset{\sim}{\beta}^0\right) = \sqrt{n}\,\underset{\sim}{Q}^{-1} \sum_{i=1}^{n} \underset{\sim}{x}'_i \varphi_{\alpha_1,\alpha_2}(e_i) + O_p\left(\frac{1}{\sqrt[4]{n}}\right),$$

where

$$\varphi_{\alpha_1,\alpha_2}(t) = \begin{cases} F^{-1}(\alpha_1) & \text{if} \quad t < F^{-1}(\alpha_1). \\ t & \text{if} \quad F^{-1}(\alpha_1) \le t \le F^{-1}(\alpha_2). \\ F^{-1}(\alpha_2) & \text{if} \quad F^{-1}(\alpha_2) < t. \end{cases}$$

This representation implies asymptotic normality of $\hat{\underset{\sim}{\beta}}^{TLS}_{\alpha_1,\alpha_2}$, i.e.,

$$(71) \quad \mathcal{L}\left(\sqrt{n}\left(\hat{\underset{\sim}{\beta}}^{TLS}_{\alpha_1,\alpha_2} - \underset{\sim}{\beta}^0_{\alpha_1,\alpha_2}\right)\right) \longrightarrow \mathcal{N}_p\left(\underset{\sim}{0},\, \sigma^2(\alpha_1,\alpha_2,F)\, \underset{\sim}{Q}^{-1}\right),$$

where

$$\sigma^2(\alpha_1,\alpha_2,F) = \frac{1}{(\alpha_1-\alpha_2)^2}\left\{\int_{\alpha_1}^{\alpha_2}\left(F^{-1}(u)-\delta\right)^2 du + \right.$$
$$+\alpha_1\left(F^{-1}(\alpha_1)-\delta\right)^2 + (1-\alpha_2)\left(F^{-1}(\alpha_2)-\delta\right)^2 -$$
$$\left. - \left[\alpha_1\left(F^{-1}(\alpha_1)-\delta\right) + (1-\alpha_2)\left(F^{-1}(\alpha_2)-\delta\right)\right]^2\right\}$$

with

$$\delta = \frac{1}{\alpha_2-\alpha_1}\int_{\alpha_1}^{\alpha_2} F^{-1}(u)\,du$$

and

$$\underset{\sim}{\beta}^0_{\alpha_1,\alpha_2} = \left(\beta_1^0+\delta,\ \beta_2^0,\ldots,\beta_p^0\right)'.$$

For all above results see Ruppert and Caroll (1980) and Jurečková (1984). From the presented results it is clear why we "prefer" symmetry of F (in practice attainable – at least approximately – by a transformation; see Carroll and Ruppert (1988)). They also suggested symmetric trimming. It is the case when we obtain unbiased estimator of intercept. This is also the reason why tests of symmetry of residuals have been constructed (Jurečková (1971), Víšek (1991)). A disadvantage of $\hat{\underset{\sim}{\beta}}_\alpha$ (as well as $\hat{\underset{\sim}{\beta}}^{TLS}$) is its sensitivity to leverage points. In 1985 Antoch and Jurečková proposed an estimator which, using the diagonal elements of the hat matrix $\underset{\sim}{X}(\underset{\sim}{X}'\underset{\sim}{X})^-\underset{\sim}{X}'$, tries to detect relevant leverage points in factor space and after rejecting them evaluates $\hat{\underset{\sim}{\beta}}^{TLS}$. In more details, let $0 < \delta < 1$ be fixed and put $\underset{\sim}{B} = \{b_{ij}\}_{i,j=1}^n$, $b_{ij} = \delta_{ij} I_{\{h_{ii}<p/n\delta\}}$. Then the *$\alpha$-trimmed least squares estimator resistant to the leverage points* is defined as

$$\hat{\underset{\sim}{\beta}}^{RTLS} = \hat{\underset{\sim}{\beta}}^{TLS}(\underset{\sim}{B}\underset{\sim}{Y},\ \underset{\sim}{B}\underset{\sim}{X}).$$

In Antoch and Jurečková (1985) the authors also found an asymptotic representation for $\hat{\underset{\sim}{\beta}}^{RTLS}$ analogous to (71) and derived a convergence in distribution similar to (72).

7k. Testing submodels.

At the end of the discussion devoted to the L-estimators let us mention briefly testing hypothesis about equivalence of the original model with some submodel. The reason is that we would like to illustrate that we can do more than point estimation, construction of (asymptotic) confident intervals and ellipsoid of the confidence, consistency or asymptotic normality etc.

So, let us assume a hypothesis

$$H_0: \beta_j = 0 \qquad j = r+1, \ldots, p, \quad 1 \le r \le p-1.$$

Let the design matrix and vector of regression coefficients be partitioned so that $\underset{\sim}{X} = ({}_r\underset{\sim}{X} \,|\, {}_q\underset{\sim}{X})$, ${}_r\underset{\sim}{\beta} = (\beta_1, \ldots, \beta_r)'$, ${}_q\underset{\sim}{\beta} = (\beta_{r+1}, \ldots, \beta_p)'$. Further let

$${}_r\underset{\sim}{\hat{\beta}}^* = \left(\left(\underset{\sim}{\hat{\beta}}^{TLS}(\underset{\sim}{Y}, {}_r\underset{\sim}{X})\right)', 0, \ldots, 0\right)'.$$

Notice that the last $p-r$ coordinates are equal to zero because $\underset{\sim}{\hat{\beta}}^{TLS}(\underset{\sim}{Y}, {}_r\underset{\sim}{X})$ has only r coordinates. Finally put

$$T_n = \frac{1}{1-2\alpha} \cdot \frac{1}{S_n^2} \left(\underset{\sim}{\hat{\beta}}^{TLS}(\underset{\sim}{Y}, \underset{\sim}{X}) - {}_r\underset{\sim}{\hat{\beta}}^*\right)(\underset{\sim}{X}'\underset{\sim}{D}\underset{\sim}{X}) \cdot \left(\underset{\sim}{\hat{\beta}}^{TLS}(\underset{\sim}{Y}, \underset{\sim}{X}) - {}_r\underset{\sim}{\hat{\beta}}^*\right),$$

where

$$S_n^2 = \frac{1}{(1-2\alpha)^2}\left[\frac{1}{n-p} \cdot (\underset{\sim}{Y}'\underset{\sim}{D}(\underset{\sim}{I} - \underset{\sim}{X}(\underset{\sim}{X}'\underset{\sim}{D}\underset{\sim}{X})^-)\underset{\sim}{D}\underset{\sim}{Y}) + \alpha(1-\alpha)\left\{\underset{\sim}{\hat{\beta}}_{\alpha,1}^{TLS}(\underset{\sim}{Y}, \underset{\sim}{X}) - \underset{\sim}{\hat{\beta}}_{1-\alpha,1}^{TLS}(\underset{\sim}{Y}, \underset{\sim}{X})\right\}\right],$$

where $\underset{\sim}{\hat{\beta}}_{\alpha,1}^{TLS}(\underset{\sim}{Y}, \underset{\sim}{X})$ denotes the first coordinate of $\underset{\sim}{\hat{\beta}}_{\alpha}^{TLS}$ etc., and $\underset{\sim}{D}$ is given in (70). Then under H_0 the statistics T_n are asymptotically distributed according to χ_q^2. It implies that one may construct a confidence ellipsoid for $\underset{\sim}{\beta}^0$ as

$$C_\alpha = \left\{ \underset{\sim}{\beta} \in R_p \,\middle|\, \frac{1}{1-2\alpha}\frac{1}{S_n{}^2}\left(\underset{\sim}{\hat{\beta}}^{TLS}(\underset{\sim}{Y}, \underset{\sim}{X}) - \underset{\sim}{\beta}\right)(\underset{\sim}{X}'\underset{\sim}{D}\underset{\sim}{X}) \cdot \left(\underset{\sim}{\hat{\beta}}^{TLS}(\underset{\sim}{Y}, \underset{\sim}{X}) - \underset{\sim}{\beta}\right) \le \chi_p^2(1-\alpha)\right\},$$

where $\chi_p^2(1-\alpha)$ is $(1-\alpha)$-quantile of χ_p^2. For details see Ruppert and Carroll (1980), Jurečková (1983), Antoch (1986) or Antoch (1987).

71. Adaptive estimators.

The challenge of obtaining efficient estimators under the assumption that we do not know the underlying parametric family has attracted the attention of many statisticians for years. For the first formulation of the problem see Stein (1956). In an analogy with Beran's solution of the problem for a location parameter we may try to define an adaptive estimator of regression coefficients. The crucial assumption of Beran's approach is symmetry of F. There are also some other assumptions, naturally, but they are of technical character. Beran's basic idea is the following one.

Let $\hat{f}_n(y, \underset{\sim}{Y}, \underset{\sim}{\beta})$ be a kernel estimator of the density cf the residuals, i. e.,

$$\hat{f}_n(y, \underset{\sim}{Y}, \underset{\sim}{\beta}) = \frac{1}{nc_n} \sum_{i=1}^{n} k\left(c_n^{-1}\left(y - (Y_i - \underset{\sim}{x}_i' \underset{\sim}{\beta})\right)\right)$$

where k is a symmetric kernel and c_n corresponding bandwidth. For $\underset{\sim}{\beta} = \underset{\sim}{\beta}^0$ the residuals are symmetrically distributed and hence $\hat{f}_n(y, \underset{\sim}{Y}, \underset{\sim}{\beta}^0)$ and

$$\frac{1}{nc_n} \sum_{i=1}^{n} k\left(c_n^{-1}\left(y + (Y_i - \underset{\sim}{x}_i' \underset{\sim}{\beta}^0)\right)\right),$$

which due to the symmetry of the kernel k is equal to $\hat{f}_n(-y, \underset{\sim}{Y}, \underset{\sim}{\beta}^0)$, should be "statistically" equal each other. So, we may define $\hat{\underset{\sim}{\beta}}^{ADA}$ as that point of R_p for which Hellinger's distance between $\hat{f}_n(y, \underset{\sim}{Y}, \underset{\sim}{\beta})$ and $\hat{f}_n(-y, \underset{\sim}{Y}, \underset{\sim}{\beta})$ is minimal. It is necessary to cope with some technical problems in the tails of the density estimator. Hence let for $\{a_n\}_{n=1}^{\infty} \nearrow \infty$ $b_n(y) = I_{[-a_n, a_n]}(y)$ and put

$$h_n(y, \underset{\sim}{Y}, \underset{\sim}{\beta}) = \sqrt{\hat{f}_n(y, \underset{\sim}{Y}, \underset{\sim}{\beta}) \cdot b_n(y)}.$$

Now, we define

$$\begin{aligned} \hat{\underset{\sim}{\beta}}^{AD_1} &= \underset{\underset{\sim}{\beta} \in R_p}{\arg\min} \int \left[h_n(y, \underset{\sim}{Y}, \underset{\sim}{\beta}) - h_n(-y, \underset{\sim}{Y}, \underset{\sim}{\beta})\right]^2 dy \\ &= \underset{\underset{\sim}{\beta} \in R_p}{\arg\max} \int h_n(y, \underset{\sim}{Y}, \underset{\sim}{\beta}) \cdot h_n(-y, \underset{\sim}{Y}, \underset{\sim}{\beta})\, dy. \end{aligned}$$

Although the estimator $\hat{\underset{\sim}{\beta}}^{AD_1}$ is consistent, it is not efficient. This follows from the asymptotic representation which has the form

$$\frac{I(f)}{\sqrt{n}} \sum_{l=1}^{p} \left(\hat{\underset{\sim}{\beta_l}}^{AD_1} - \underset{\sim}{\beta_l}^{0} \right) \sum_{i=1}^{n} x_{il} =$$

$$= \frac{1}{\sqrt{n}} \sum_{i=1}^{n} \{ f' (Y_i - \underset{\sim}{x}_i' \underset{\sim}{\beta}^0) \cdot f^{-1} (Y_i - \underset{\sim}{x}_i' \underset{\sim}{\beta}^0) \} + o_p(1),$$

where $I(f)$ is the Fisher information of f. Notice that the last relation reduces to the analogous Beran's estimator for location parameter.

In spite of a strange asymptotic representation of this estimator it turns out that for practical purposes it behaves quite well giving "the most symmetrical" residuals among all methods. Hence in a situation when the physical circumstances (strongly) support the assumption of symmetry, it may be useful to employ this estimator.

The lack of efficiency of this estimator has led to the idea to use the kernel estimator of residual density in another way for estimating regression coefficients. Let us assume for a while that we know f precisely. Then we may use the MLE of the regression coefficients, namely,

$$\hat{\underset{\sim}{\beta}}^{MLE} = \arg\max_{\underset{\sim}{\beta} \in R_p} \sum_{i=1}^{n} \log f (Y_i - \underset{\sim}{x}_i' \underset{\sim}{\beta}).$$

So, the straightforward idea is to use a preliminary (consistent and robust) estimator $\hat{\underset{\sim}{\beta}}^*$ of regression coefficients, establish $\hat{f}(y, \underset{\sim}{Y}, \hat{\underset{\sim}{\beta}}^*)$ and define

$$\hat{\underset{\sim}{\beta}}^{AD_2} = \arg\max_{\underset{\sim}{\beta} \in R_p} \sum_{i=1}^{n} \log \hat{f} \left(Y_i - \underset{\sim}{x}_i' \underset{\sim}{\beta}, \underset{\sim}{Y}, \hat{\underset{\sim}{\beta}}^* \right) \cdot b_n (Y_i - \underset{\sim}{x}_i' \underset{\sim}{\beta}).$$

Such estimators are, provided F is symmetrical and some other technical assumptions hold (for details see Víšek (1991)), consistent, asymptotically normal and efficient.

Another way to construct adaptive estimators has been proposed by Dodge and Jurečková (1987) and studied later by Antoch, Dodge and Jurečková (1991). They showed that the estimation of the linear regression parameter vector by a convex combination of the least squares and the least absolute deviations estimators could be adaptive, so that the resulting estimator achieves the minimum asymptotic variance in the model under consideration. Let us shortly summarize their idea.

We will consider again the model $\underset{\sim}{Y} = \underset{\sim}{X}\,\underset{\sim}{\beta} + \underset{\sim}{e}$ and assume moreover, that errors are i. i. d. according to the density $f(x)$ satisfying

(72) $$f(x) = f(-x), \quad \forall x \in R_1, \qquad 0 < f(0) < +\infty$$

and

$$0 < \sigma^2 = \int_{R_1} x^2 f(x)\,dx < +\infty,$$

$f(x)$ has a bounded derivative in a neighborhood of 0 and

$$f(x) = \frac{1}{s} f^0\left(\frac{x}{s}\right), \qquad s > 0,\ x \in R_1,$$

where $f^0(x)$ is a fixed, generally unknown density satisfying (72) and normalized so that $f^0(0) = 1$.

Let $\hat{\underset{\sim}{\beta}}^{L_1}$ denote the L_1-norm estimator. Fix $\varepsilon > 0$ and put

(73) $$V_n(\varepsilon) = \frac{1}{2\varepsilon\sqrt{n}} \sum_{i=1}^{n} I\left[\left|Y_i - \underset{\sim}{x}_i'\hat{\underset{\sim}{\beta}}^{L_1}\right| \le \frac{\varepsilon}{\sqrt{n}}\right],$$

where $I[\cdot]$ denotes the indicator function. It can be shown that $V_n^{-1}(\varepsilon)$ is a consistent estimator of σ^2. Further, put

(74) $$\hat{E}_1^0 = \frac{V_n(\varepsilon)}{n-p} \sum_{i=1}^{n} \left|Y_i - \underset{\sim}{x}_i'\hat{\underset{\sim}{\beta}}^{L_1}\right|$$

and

(75) $$\hat{\sigma}_0^2 = \frac{V_n^2(\varepsilon)}{n-p} \sum_{i=1}^{n} \left(Y_i - \underset{\sim}{x}_i'\hat{\underset{\sim}{\beta}}^{L_1}\right)^2.$$

The Dodge and Jurečková adaptive estimator $\hat{\underset{\sim}{\beta}}^{DJ}$ of $\underset{\sim}{\beta}$ is defined as follows:

A1. Calculate $\hat{\underset{\sim}{\beta}}^{L_1}$.

A2. Fix $\varepsilon > 0$ and compute $V_n(\varepsilon)$, $\hat{E}_1^0$ and $\hat{\sigma}_0^2$ according to (73)–(75).

A3. If $\hat{E}_1^0 \ge 1/2$, use the L_1-norm estimator as $\hat{\underset{\sim}{\beta}}^{DJ}$.

A4. If $\hat{E}_1^0 < 1/2$ and $\hat{E}_1^0 \ge 2\hat{\sigma}_0^2$, use the classical least squares estimator as $\hat{\underset{\sim}{\beta}}^{DJ}$.

A5. If $\hat{E}_1^0 < 1/2$ and $\hat{E}_1^0 < 2\hat{\sigma}_0^2$, calculate

$$\hat{\delta}_0 = \frac{4\hat{\sigma}_0^2 - 2\hat{E}_1^0}{4\hat{\sigma}_0^2 - 4\hat{E}_1^0 + 1} \tag{76}$$

and use as $\hat{\underset{\sim}{\beta}}^{DJ}$ the solution of the minimization problem

$$\arg\min_{\underset{\sim}{t} \in R_p} \sum_{i=1}^{n} \rho_\delta((Y_i - \underset{\sim}{x}_i' \underset{\sim}{t}) \cdot V_n(\varepsilon)), \tag{77}$$

where $\delta = \hat{\delta}_0$ and

$$\rho_\delta(x) = (1-\delta)x^2 + \delta|x|, \qquad x \in R_1. \tag{78}$$

In practice, of course, we must replace δ_0 by an appropriate estimator based on the vector of observations $\underset{\sim}{Y}$.

Intuitively speaking, based on the preliminary estimator $\hat{\underset{\sim}{\beta}}^{L_1}$, the method tries to offer the "optimal" solution from the range going from the L_1 to the L_2 norm estimator. If the distribution of errors has "heavy tails", i. e., tails heavier than those of the double exponential distribution, the method will use the L_1-norm estimator. On the other hand, if the tails of F are not heavier than those of the normal distribution, the resulting estimator will be the classical least squares one. Finally, if the tails of F lie between those of the normal and the double exponential distributions, the estimator used will be the M-estimator generated through (77) and (78). Let us point out once more that the resulting estimator achieves the minimum asymptotic variance in the model under consideration. The asymptotic properties of the proposed estimator are studied in more details in the paper of Jurečková and Dodge (1987).

7m. Computation of the L-estimators.

We could see in section 7i. that the regression quantiles play the key role in definition of the L-estimators in the regression model. For their calculation the following three approaches can be recommended.

i) One can take into account that the γ-regression quantile is in fact the M-estimator generated by the ρ-function of the form $\rho(x) = x\ (\gamma - I_{[x<0]})$ and use any available method for the evaluation of M-estimators. For example we can put $w(x) =$

$\dfrac{\gamma - I_{[x<0]}}{x}$, where $I[x]$ denotes the indicator function, and use *IRLS* starting from *LS* solution. However, the convergence is usually rather slow even for moderate sample sizes.

ii) The second approach is quite different. It is based on the representation of the regression quantiles as the solution of the linear programming problem (69), which can be easily solved using the simplex method. Moreover, due to the very special form of the objective function and the simple constraints, several modifications of the classical simplex have been proposed and their properties not only studied, but also implemented on computers. These methods usually pass through several vertices during one iteration step, what enables to find the solution usually in a very few steps. But what is even more appreciable on this approach is the fact that no preliminary estimator of β is necessary. For more details see papers of Barrodale and Roberts (1973, 1974 and 1977) among others.

iii) The third approach uses the dual problem corresponding to the linear programming representation. For more details see Koenker and D'Orey (1987).

It is evident that the rest of the problem is very simple. In fact, it is nothing else than the routine use of the classical least squares method which follows directly the definition of the L-estimators.

7n. Use of the iteratively reweighted least squares method for the computation of M-estimators in linear models.

We have already pointed out that the simplest way to compute M-estimators is the so called iteratively reweighted least squares (*IRLS*) method. Because this approach is extremely simple to implement, we will describe it more precisely.

We will again consider the model $\underset{\sim}{Y} = \underset{\sim}{X}\ \underset{\sim}{\beta} + \underset{\sim}{e}$ described in more detail earlier. Moreover, let $\rho(\underset{\sim}{x}, \underset{\sim}{\beta})$ denotes the ρ-function generating the desired M-estimator. Our aim is to find that $\hat{\underset{\sim}{\beta}}$ which minimizes

$$\sum_{i=1}^{n} \rho\left(\frac{Y_i - \sum_{j=1}^{p} x_{ij}\beta_j}{S_n}\right), \tag{79}$$

where $S_n > 0$ is a fixed known constant used as an estimator of the scale parameter. Several possible statistics S_n have been discussed in Section 3.

It has been shown earlier that in the case when $\psi(\equiv \rho')$ exists, the solution of the minimization problem (79) is equivalent to the solution of the system of equations

$$\sum_{i=1}^{n} x_{ij}\psi\left(\frac{Y_i - \sum_{j=1}^{p} x_{ij}\beta_j}{S_n}\right) = 0, \qquad j = 1,\dots,p. \tag{80}$$

Let us denote $w(x) = \psi(x)/x$ for $x \neq 0$. Then using elementary algebraic operations we can rewrite (80) in the following form, i. e., that for $j = 1,\dots,p$

$$\sum_{i=1}^{n} x_{ij}w\left(\frac{Y_i - \sum_{j=1}^{p} x_{ij}\beta_j}{S_n}\right)\left(Y_i - \sum_{k=1}^{p} x_{ik}\beta_k\right) = 0. \tag{81}$$

Let us denote

$$\underset{\sim}{V} = \underset{\sim}{V}(\underset{\sim}{\beta}) = \text{diag}\left(w\left(\frac{e_1(\underset{\sim}{\beta})}{S_n}\right),\dots,w\left(\frac{e_n(\underset{\sim}{\beta})}{S_n}\right)\right), \tag{82}$$

where $e_i(\underset{\sim}{\beta}) = Y_i - \sum_{j=1}^{p} x_{ij}\beta_j$ denotes the $i-$th residual corresponding to the current value of $\underset{\sim}{\beta}$. Moreover, let $\underset{\sim}{W} = \underset{\sim}{W}(\underset{\sim}{\beta}) = \sqrt{\underset{\sim}{V}}$. Then it is evident that using the matrix representation we can rewrite (3) in the form

$$\underset{\sim}{X}'\underset{\sim}{V}\underset{\sim}{X}\underset{\sim}{\beta} = \underset{\sim}{X}'\underset{\sim}{V}\underset{\sim}{Y}.$$

But this is nothing else than the system of normal equations for the following weighted least squares problem, i.e., to minimize with respect to $\underset{\sim}{\beta}$

$$\sum_{i=1}^{n}\left(w\left(\frac{e_i(\underset{\sim}{\beta})}{S_n}\right)Y_i - \sum_{j=1}^{p} x_{ij}w\left(\frac{e_i(\underset{\sim}{\beta})}{S_n}\right)\beta_j\right)^2. \tag{83}$$

In other words, (81) is nothing else than the system of normal equations for the weighted least squares problem in which the weights depend, through the residuals, on the current value of $\underset{\sim}{\beta}$.

From (82) and (83) it is already evident how to proceed further. Let us denote $\hat{\underset{\sim}{\beta}}^{(0)}$ any initial estimator of $\underset{\sim}{\beta}$. Then desired solution is given by the following simple iterative schema, i.e.,

$$\underset{\sim}{X}'\underset{\sim}{V}(\hat{\underset{\sim}{\beta}}^{(k-1)})\underset{\sim}{X}\,\underset{\sim}{\beta} = \underset{\sim}{X}'\underset{\sim}{V}(\hat{\underset{\sim}{\beta}}^{(k-1)})\underset{\sim}{Y}, \qquad k = 1,2,\dots \tag{84}$$

It is evident that in (84) $\hat{\underset{\sim}{\beta}}^{(k-1)}$ denotes the solution from the previous step, $\underset{\sim}{V}(\hat{\underset{\sim}{\beta}}^{(k-1)})$ are the weights corresponding to this solution and $\underset{\sim}{\beta}$ is the vector of unknown parameters which we are estimating in the k-th step, $k = 1, 2, \ldots$. The name of this method originates from the fact that the classical least squares solution is used each step with the weights iteratively recomputed in each step based on the previous solution.

As to the initial solution $\hat{\underset{\sim}{\beta}}^{(0)}$ one can take any of the previously discussed estimators. Nevertheless, the most commonly used one in the practice is the L_p-norm solution, usually with $p = 2$ or $p = 1$, which correspond to the least squares and least absolute deviations estimators. It is clear, that in the presence of highly influential points these initial estimators can be very bad.

Another important question is when to stop the iterative process. The most commonly used approach is to stop if two consecutive solutions, say $\hat{\underset{\sim}{\beta}}^{(l)}$ and $\hat{\underset{\sim}{\beta}}^{(l+1)}$, are close enough, i. e., if $\left\| \hat{\underset{\sim}{\beta}}^{(l+1)} - \hat{\underset{\sim}{\beta}}^{(l)} \right\| < \kappa$, κ being a fixed small positive constant set in advance. Nevertheless, in any case one should try the solution in (81) to assess its preciseness.

For the case when we want to estimate both $\underset{\sim}{\beta}$ and the scale parameter, the *IRLS* algorithm must be modified as follows. We start from any initial estimators $\hat{\underset{\sim}{\beta}}^{(0)}$ and $\hat{S}_n^{(0)}$. Then we find $\hat{\underset{\sim}{\beta}}^{(1)}$ from (84) and recalculate $\hat{S}_n^{(1)}$ using the residuals corresponding to $\hat{\underset{\sim}{\beta}}^{(1)}$ etc. The iterative process is stopped, e. g., if two consecutive solutions are close enough.

There exist several different modifications of the *IRLS* method. Most of them have been proposed and studied by the specialists from the field of numerical analysis. One of the most up to date comparison among them can be found in Wolke (1991). In this paragraph we have described only the simplest variant of the *IRLS* approach (due to lack of space). Nevertheless, according to our experience it works – and we can insist that it works well in most of the practical situations.

Another important point is that the *IRLS* method can be used for the computation of many other estimators, too. Not only those more or less deeply connected with the M-estimators, but also for the computation of R- and L-estimators etc. For more details see Hetmansperger (1985) or Antoch (1985) among others.

7o. One (and k) step estimators.

At the end of our discussion about robust regression analysis let us turn our attention to a few topics which may be also of interest. As we have noticed earlier, one way to improve the behavior of the estimators of regression coefficients is to "recalculate" them iteratively. We may start with some "very" robust estimator, let us say with high breakdown point although with low efficiency, and then evaluate one-step (or k-step) estimators employing in some way Gauss-Newton iteration scheme. It has already been observed by R. A. Fisher, J. von Neuman and others for the invariant parameter, that in the case when we know F precisely, put $\psi = f'/f$ and start with any $\sqrt{n}$ consistent estimator $\hat{\beta}^0$, then when performing one Gauss-Newton iteration step in the equation

$$\sum_{i=1}^{n} \psi\left(\alpha_i - \hat{\beta}^0\right) = 0,$$

we obtain estimator which is asymptotically efficient, even when the MLE is not, and is equivalent to MLE when this is. Peter J. Bickel in 1975 generalized this assertion on regression scheme. The simplest version of his result may be given for absolutely continuous ψ with derivative ψ'. Suppose that $X'X$ is regular, then $\hat{\beta}^{OS}$ is given by

$$\hat{\beta}^{OS} = \hat{\beta}^0 + \tag{85}$$

$$+ \left[\hat{A}(\psi, F)\right]^{-1} \left(\psi\left(Y_1 - x_1'\hat{\beta}^0\right), \ldots, \psi\left(Y_n - x_n'\hat{\beta}^0\right)\right) X[X'X]^{-1},$$

where $\hat{A}(\psi, F)$ is an estimator of

$$A(\psi, F) = \int_{-\infty}^{\infty} \psi(t)\, dF(t).$$

Later on Jurečková and Portnoy (1987) developed asymptotics of one-step M-estimators. They have proved that even if the initial estimator $\hat{\beta}^0$ is less than $\sqrt{n}$ consistent, we may obtain $\sqrt{n}$ consistency and high efficiency for $\hat{\beta}^{OS}$, which also inherits the level of breakdown point from $\hat{\beta}^0$.

Although the theory is not simple, the heuristic behind it is quite transparent. It is sufficient to realize that properties like "break down point" are determined by the "finite sample behavior" of the estimator. Don't be confused that we (usually) speak about asymptotic value of breakdown point. Realize that – at least for robust estimators with high breakdown point – we always evaluate the breakdown point for finite samples and then we take the limit for sample size going to infinity. A little different and somewhat more complicated argument leads to the same effect for the "theoretical" breakdown point originally introduced by Hampel (1968). And the finite sample behavior of $\hat{\underset{\sim}{\beta}}^{OS}$ is mainly determined by $\hat{\underset{\sim}{\beta}}^{0}$.

On the other hand, the asymptotic features of estimation are given by behavior "infinity", i. e. roughly speaking it may happen that all finite features of $\hat{\underset{\sim}{\beta}}^{0}$ may asymptotically "disappear". Notice that for instance when studying efficiency you take at first limit (in distribution) and then you look at the variance of the asymptotic distribution; not the limit of var $\hat{\underset{\sim}{\beta}}_n^{OS}$, if any!

Nowadays k–step estimators are studied and results show that in some cases for practical purposes, especially when starting from a not very robust initial estimator, it is necessary to perform even hundreds of steps (we will discuss it later on).

7p. Tail behavior.

Another interesting result is due to He, Jurečková, Koenker and Portnoy (1990) and concerns the finite behavior of regression estimators. The result is a generalization of earlier results of Jurečková (1981) and Antoch (1984). He et al. studied the behavior of

$$B\left(z, \hat{\underset{\sim}{\beta}}_n\right) = \frac{-\log P\left\{\max_{1\le i\le n}\left|\underset{\sim}{x}_i'\left(\hat{\underset{\sim}{\beta}}_n - \underset{\sim}{\beta}^0\right)\right| > z\right\}}{-\log(1-F(z))}$$

for $z \to \infty$. To be able to present their result let us define a class of so called light – tailed distributions. A random variable Z with the distribution function $F(z)$ will belong to this class if for some $r > 0$ and $c > 0$

$$\lim_{z\to\infty} \frac{-\log(1-F(z))}{cz^r} = 1.$$

On the contrary, we will assign the random variable Z with d. f. $F(z)$ the heavy tailed one if

$$\lim_{z\to\infty} \frac{-\log(1-F(z))}{m\log z} = 1$$

for some $m > 0$. For the light – tailed distribution and for $1 < r \le 2$ we have for $\hat{\underset{\sim}{\beta}}^{LS}$

$$\left(\overline{h}\right)^{1-r} \le \liminf_{z\to\infty} B\left(z, \hat{\underset{\sim}{\beta}}^{LS}\right) \le \limsup_{z\to\infty} B\left(z, \hat{\underset{\sim}{\beta}}^{LS}\right) \le \left(\overline{h}\right)^{-r},$$

where $\overline{h} = \max_{1 \le i \le n} \{ \underset{\sim}{X}_i'(\underset{\sim}{X}'\underset{\sim}{X})^{-1}\underset{\sim}{X}_i \}$. For $r = 1$ then

$$\left(\overline{h}\right)^{-1/2} \le \liminf_{z \to \infty} B\left(z, \underset{\sim}{\hat{\beta}}^{LS}\right) \le \limsup_{z \to \infty} B\left(z, \underset{\sim}{\hat{\beta}}^{LS}\right) \le \left(\overline{h}\right)^{-1}.$$

For the normal distribution one obtains

$$\lim_{z \to \infty} B\left(z, \underset{\sim}{\hat{\beta}}^{LS}\right) = \left(\overline{h}\right)^{-1}$$

and for the heavy-tailed distribution finally

$$\lim_{z \to \infty} B\left(z, \underset{\sim}{\hat{\beta}}^{LS}\right) = 1.$$

The same authors also derived analogous results for L- and M-estimators, but they are still in a little bit "raw" form and it would take more space to prepare the framework for their presentation. Let us restrict ourselves on that for LMS-estimator. They are bounded below by:

$$\liminf_{z \to \infty} B\left(z, \underset{\sim}{\hat{\beta}}^{LMS}\right) \ge \left[\frac{n}{2}\right] - p + 2.$$

All results hint that there is a connection (not yet fully understable) between breakdown point and tail behavior of estimators.

8. Some concluding remarks

As follows from this contribution, robust statistics offers today a very broad scale of methods for location, scale and regression estimation. Unfortunately, software for them is far from being generally available. On the other hand, robust methods have started to make their way into statistical packages. Moreover, as it was explained above, for some robust methods it is not difficult to write macros using standard ("classic") packages, usually employing subroutines for the LSE. Then the question is what robust method should be used. It is nearly sure that no method can cover all problems especially when processing "real" data. Although this fact is commonly accepted, a lot of people use one "favorite" method, usually LS or L_1, or alternatively some of them look for a new miraculous method, preferably supported by "objective" laws of nature, which would break less often.

Since the present theory of robust estimation (and testing) brings all the time weaker and weaker (sometimes even necessary) conditions under which such properties as consistency, asymptotic normality etc.

hold, it is clear that further theoretical study will eventually offer criteria to distinguish among methods and to describe more completely their behavior under presence of contamination of various types.

It seems also unquestionable that further study(ies) of Princeton's type would be more than welcomed. However, for what concerns regression analysis, due to the fact that it is not a mere generalization of the location problem, one can hardly hope for too many exhaustive numerical studies.

Previous explanation also hinted that in contrast with classical statistics, results of robust methods need much more careful interpretation. Taking moreover into account a broad variety of methods, it is clear that some skill and experience are necessary. At a conference about the future of statistics in the early seventies Herbert Robins said: "It is not so far in history when an ill man was not advised to visit a physician. I hesitate if today a man/woman demanding to process data is aware that he/she should call for a statistician."

So, what to advise for present use of robust statistics? Do we really need robust methods, especially with high breakdown point? For the positive answer to the second question it is sufficient to realize that we never know the contamination level of data and hence we should start analysis of data by a "highly" robust method. One may object that it should be equivalent to starting with a "low" robust methods and it should turn out that they are not able to give an "admissible" solution. But it is easy to give an example of data, especially multidimensional ones, for which e. g. the LS method find "nice" regression model, "nice" in the sense of high value of the coefficient of determination as well as of acceptable values of the corresponding $t-$statistics for its coefficients, however the model will be completely wrong. The word "wrong" means that you either know the "theoretical" model, for simulated data, or for real data, you find by a method with high breakdown point a different model which yields a (much) smaller robust estimate for the variance (say $\hat{\sigma}_{MAD}$ or interquartile of residuals) and which may even throw a suspicion on some points, which may later prove to be really questionable (from physical reasons).

So, today a commonly accepted "advice" for statistical data analysis is: Run two or three procedures including one with rather high breakdown point and compare results (if possible, use a statistical test for the significance of the apparent difference the results). In the case of (significantly) different results use the result of method with high breakdown point as a diagnostics and try to "discover" physical background "causing contamination". The practitioners, especially technically educated, prefer to use LS applied on restricted sample method, may be, if it is "advised by diagnostic results", on a restricted set of data. In such a case result of LS should be such that they are not too far from the result of "appropriately robust" method. Unfortunately, the sense of "appropriately robust" is still vague and will really need a lot of effort in the future to obtain a more precise (mathematical) form. Up to now, due to this lack of theoretical results, the selection of an "appropriately robust" method is left to the experience of a statistician (for some first result see Víšek (1987), (1989) and Rubio de Juan A., Víšek (1991)).

The advantage of robust methods, namely their resistance to contamination, guarantees that the idea of robustness – mathematically expressed in families of approximate models of data – will spread successively into all parts of statistics. If the present report brings you to the point that it is useful (or even necessary) to think about it once again it has fulfilled its goal. Good luck when processing data!

References

[1] Adelmalek N. N., *On the discrete ℓ_1 – approximation and ℓ_1 – solutions of overdetermined linear equations*, Journal of Approximation Theory **11** (1974), 38 – 53.

[2] Andrews D. F. et al., *Robust estimates of location. Survey and advances*, Princeton University Press, Princeton, 1972.

[3] Andrews D. F. and Pregibon D., *Finding outliers, that matter*, Journal of the Royal Statistical Society **B 40** (1978), 85 – 93.

[4] Antoch J., *Behavior of L – estimators of location from the point of view of large deviations*, In: Robustness of statistical methods and non – parametric statistics (Rasch D. and Tiku M. L., eds.), VEB, Berlin, 1984, pp. 5 – 8.

[5] Antoch J., *Numerical behavior of L – estimators in general linear model*, Computational Statistics Quarterly **2** (1985), 363 – 381.

[6] Antoch J., *Algorithmic development in variable selection procedures*, In: Proceedings of COMPSTAT'86 (Rizzi A. et al., eds.), Physica – Verlag, Wien, 1986, pp. 83 – 90.

[7] Antoch J., Collomb G. and Hassani S., *Robustness in parametric and nonparametric regression estimation*, In: Proceedings of COMPSTAT'84 (Havránek T. et al., eds.), Physica – Verlag, Wien, 1984, pp. 49 – 54.

[8] Antoch J. and Jurečková J., *Trimmed least squares estimator resistant to leverage points*, Computational Statistics Quarterly **2** (1985), 329 – 339.

[9] Armstrong R. D., *A revised simplex algorithm for the absolute deviation curve fitting problem*, Communications in Statistics – Simulation Computation **B8** (1979), 175 – 190.

[10] Atkinson A. C., *Plots, transformations and regression*, Oxford University Press, New York, 1985.

[11] Barnett V. and Lewis T., *Outliers in statistical data*, J. Wiley, New York, 1977.

[12] Barnett V. and Lewis T., *Outliers in statistical data (2nd edition)*, J. Wiley, New York, 1985.

[13] Barrodale I. and Roberts F. D. K., *An improved algorithm for discrete ℓ_1 − norm approximation*, SIAM Journal of Numerical Analysis **10** (1973), 839-849.

[14] Barrodale I. and Roberts F. D. K., *Algorithms for restricted least absolute value estimation*, Communications in Statistics – Simulation Computation **B6** (1977), 353 – 363.

[15] Bassett G. and Koenker R., *Asymptotic theory of least absolute error regression*, Journal of the American Statistical Association **77** (1978), 407 – 415.

[16] Belsley D. A. et al., *Regression diagnostics. Identifying influential data and sources of colinearity*, J. Wiley, New York, 1980.

[17] Beran R., *An efficient and robust adaptive estimator of location*, Annals of Statistics **6** (1978), 292 – 313.

[18] Bickel P. J., *On some analogues to linear combinations of order statistics in the linear model*, Annals of Statistics **1** (1973), 597 – 616.

[19] Bickel P. J., *One – step Huber estimates in the linear model*, Journal of the American Statistical Association **70** (1978), 428 – 433.

[20] Bickel P. J., *Quelques aspects de la statistique robuste*, Lecture Notes in Mathematics, vol. 876, Springer, Berlin, 1981.

[21] Bloomfield P. and Steiger W. L., *Least absolute deviations. Theory, applications and algorithms*, Birkhäser, Boston, 1983.

[22] BMDP, Distributor:, BMDP Statistical Software, Cork Technology Park, Model Farm Road, Cork, IRELAND..

[23] Carroll R. J. and Ruppert D., *Transformation and weighting in regression*, Chapman & Hall, New York, 1988.

[24] Chatterjee S. and Hadi A. S., *Sensitivity analysis in linear regression*, J. Wiley, New York, 1988.

[25] Collins J. R., *Robust estimation of a location parameter in the presence of asymmetry*, Annals of Statistics **4** (1976), 68 – 85.

[26] Cook R. D. and Weisberg S., *Residuals and influence in regression*, Chapman & Hall, London, 1982.

[27] Csörgö M. and Révész P., *Strong Approximations in Probability and Statistics*, Akademiai Kiadó, Budapest, 1981.

[28] Dodge Y. (ed.), *Statistical data analysis based on the ℓ_1 − norm and related methods*, North Holland, Amsterdam, 1987.

[29] Dodge Y., Antoch J. and Jurečková J., *Computational aspects of adaptive combination of least squares and least absolute deviations estimators*, Computational Statistics and Data Analysis **12** (1991), 87 – 99.

[30] Dodge Y. and Jurečková J., *Adaptive combination of least squares and least absolute deviations estimators*, In: Statistical data analysis based on ℓ_1 − norm and related methods (Dodge Y., ed.), Amsterdam, 1987, pp. 275 – 284.

[31] Dodge Y. and Jurečková J., *Flexible L − estimation in the linear model*, Computational Statistics and Data Analysis **12** (1991), 211 – 220.

[32] Donoho D. L. and Huber P. J., *The notion of breakdown point*, In: Festschrift for E. L. Lehmann (Bickel P. J. et al., eds.), Wadsworth, California, 1983, pp. 157 – 184.

[33] Dutter R., *An APL – function library for iteratively solving the problem of robust and bounded influence regression*, In: Proceedings of COMPSTAT 1988 (Edwards D. and Raun N. E., eds.), Physica - Verlag, Wien, 1988, pp. 219 - 226.

[34] Dutter R. and Huber P. J., *Numerical methods for the non-linear robust regression problem*, Communications in Statistics-Simulation Computation **13** (1981), 79-114.

[35] Ekblom H., *A new algorithm for the Huber estimator in linear models*, BIT **28** (1988), 123-132.

[36] Ekblom H. and Madsen K., *Algorithms for non-linear Huber estimation*, BIT **29** (1989), 60-76.

[37] Fernholz L. T., *Von Mises calculus for statistical functionals*, Lecture Notes in Statistics, vol. 19, Springer, Berlin, 1983.

[38] Field C. A. and Ronchetti E. M., *Small sample asymptotics*, Lecture Notes, Monograph Series, vol. 13, IMS, 1990.

[39] Fisher R. A., *On the mathematical foundations of theoretical statistics*, Philosophical Transactions of Royal Society London **A 222** (1922), 309-368.

[40] Franke J. et al. (eds.), *Robust and non-linear time series analysis*, Lecture Notes in Statistics, vol. 26, Springer, Berlin, 1984.

[41] Hampel F. R., *Contribution to the theory of robust estimation*, PhD thesis, University of California, Berkeley, 1968.

[42] Hampel F. R., *A general qualitative definition of robustness*, Annals of Mathematical Statistics **42** (1971), 1887-1896.

[43] Hampel F. R., *The influence curve and its role in robust estimation*, Journal of the American Statistical Association **69** (1974), 383-393.

[44] Hampel F. R. et al., *Robust statistics. The approach based on influence functions*, J. Wiley, New York, 1986.

[45] He X., Jurečková J., Koenker R. and Portnoy S., *Tail behavior of regression estimators and their breakdown points*, Econometrica **58** (1990), 1195-1214.

[46] Hetmansperger T. P., *Statistical inference based on ranks*, J. Wiley, New York, 1985.

[47] Hoaglin D. C. et al., *Understanding robust and exploratory data analysis*, J. Wiley, New York, 1983.

[48] Huber P. J., *Robust estimation of a location parameter*, Annals of Mathematical Statistics **35** (1964), 73-101.

[49] Huber P. J., *The behavior of maximum likelihood estimates under nonstandard conditions*, In: Proceedings of the 5th Berkeley symposium on mathematical statististics and probability, vol. 1, University of California Press, California, Berkeley, 1967, pp. 221-233.

[50] Huber P. J., *Robust Statistical Procedures*, Regional Conference Series in Applied Mathematics, vol. 27, SIAM, Philadelphia, 1977.

[51] Huber P. J., *Robust statistics*, J. Wiley, New York, 1981.

[52] Huber P. J., *Minimax aspects of bounded-influence regression (with discussion)*, Journal of the American Statistical Association **78** (1983), 66-80.

[53] Huber P. K. and Strassen V., *Minimax tests and Neyman-Pearson lemma for capacities*, Annals of Statististics **1** (1973), 251-263.

[54] ISP, Distributor:, DATAVISION AG, POBox471, CH-7250 Klosters, SWISS.

[55] Johns M. V., *Robust Pitman-like estimators*, In: Robustness in statistics (Launr R. L. and Wilkinson G. N., eds.), Academic Press, New York, 1979, pp. 49-60.

[56] Joss J. and Marazzi A., *Probabilistic algorithms for least median squares regression*, Computational Statistics and Data Analysis **11** (1990), 123-134.

[57] Jurečková J., *Asymptotic independence of rank test statistics for testing symmetry in regression*, Sankhya **A 33** (1971), 1 – 18.

[58] Jurečková J., *Tail behavior of location estimators*, Annals of Statististics **9** (1981), 578 – 585.

[59] Jurečková J., *Regression quantiles and trimmed least squares estimator under a general design*, Kybernetika **20** (1984), 345 – 357.

[60] Jurečková J., *M – , L – and R – estimators*, In: Handbook of statistics (Krishnaiah P.R. and Sen P.K., eds.), vol. 4, North Holland, Amsterdam, 1984, pp. 463 – 485.

[61] Jurečková J., *Consistency of M – estimators of vector parameters*, In: Proceedings of the 4th Prague symposium on asymptotic statistics (Hušková M. and Mandl P., eds.), Charles University Press, Praha, 1988, pp. 305 – 312.

[62] Jurečková J. and Portnoy S., *Asymptotics for one – step M – estimators in regression with application to combining efficiency and high breakdown point*, Communications in Statististics – Theory Methodology **A 16** (1987), 267 - 279.

[63] Jurečková J. and Sen P.K., *On adaptive scale – equivariant M – estimators in linear models*, Statistics and Decisions, Supplement Issue **1** (1984), 31 – 46.

[64] Koenker R. and Bassett G., *Regression quantiles*, Econometrica **46** (1978), 33 – 50.

[65] Koenker R.W. and Bassett G., *Tests of linear hypotheses and l_1 – estimation*, Econometrica **50** (1982), 1577 – 1583.

[66] Koenker R.W. and D'Orey V., *Computing regression quantiles*, Applied Statistics **36** (1987), 383 – 393.

[67] Lopuhaä H. P., *Estimation of location and covariance with high breakdown point*, PhD thesis, 1990.

[68] Malý M., *K – step M – estimators*, PhD thesis, Charles University, Praha, 1991.

[69] Marazzi A., *ROBETH – ROBSYS: A software for robust statistical computing*, Lausanne University Press, Lausanne, 1986.

[70] Marazzi A., *Solving bounded influence regression problems with ROBSYS*, In: Statistical data analysis based on ℓ_1 – norm (Y. Dodge ed.), North Holland, Amsterdam, 1987, pp. 145 – 161.

[71] Maronna R.A., *Robust M – estimators of multivariate location and scatter*, Annals of Statistics **4** (1976), 51 – 67.

[72] Maronna R.A. and Yohai V.I., *Asymptotic behavior of general M – estimates for regression and scale with random carriers*, Zeitschrift für Wahrscheinslichtskeitstheorie und verw. Gebiete **58** (1981), 7 – 20.

[73] Martin R.D. and Yohai V.J., *Influence functionals for time series*, Annals of Statistics **14** (1986), 781 – 855.

[74] Narula S.C. and Wellington J.F., *The minimum sum of absolute errors regression. A state of the art survey*, International Statistical Review **50** (1982), 317 – 326.

[75] Peters U. and Willms C., *Program for computing linear regression in ℓ_1 – norm with downdating and updating for rows and columns*, Operation Research Spectrum **5** (1983), 229 – 239.

[76] Rey W.J.J., *Introduction to robust and quasi – robust statistical methods*, Lecture Notes in Mathematics, vol. 690, Springer, Berlin, 1983.

[77] Rousseeuw P.J., *Least median of squares regression*, Journal of the American Statistical Association **79** (1984), 871 – 880.

[78] Rousseeuw P. J. and Leroy A. M., *Robust regression and outlier detection*, J. Wiley, New York, 1987.

[79] Ruppert D. and Carroll R. J., *Trimmed least squares estimation in the linear model*, Journal of the American Statistical Association **75** (1980), 828 – 838.

[80] S+, Distributor:, Statistical Sciences Inc., 1700 Westlake Avenue North, Suite 500, Seatle, WA 98109, USA.

[81] Sallas M. W., *An algorithm for an ℓ_p – norm fit of a multiple linear regression model*, IMSL Technical Report, IMSL, Houston, 1990.

[82] Samarov A. M. and Welsh R. E., *Computational procedures for bounded influence regression*, In: Proceedings of COMPSTAT'82 (Caussinus H. P. et al., eds.), Physica – Verlag, Wien, 1982, pp. 244 – 248.

[83] SAS, Distributor:, SAS Institute Inc., SAS Campus Drive, Cary, NC 27513, USA.

[84] Sen A. and Srivastava M., *Regression analysis. Theory, methods and applications*, Springer, New York, 1990.

[85] Serfling R. J., *Approximation theorems of mathematical statistics*, J. Wiley, New York, 1980.

[86] Siegel A. F., *Robust regression using repeated medians*, Biometrika **69** (1982), 242 – 244.

[87] Späth H., *Mathematische Software zür Lineare Regression*, Oldenburg Verlag, München, 1987.

[88] Sposito V., Smith N. and McCormick G., *Minimizing the sum of absolute deviations*, Vanderhoek & Ruppert, Götingen, 1978.

[89] SPSS, Distributor:, SPSS Inc., 444 N Michigan Avenue, Chicago, IL 60611, USA.

[90] Stahel W. and Weisberg S., *Directions in robust statistics I. & II.*, The IMA values in mathematics and its applications, Vol. 33 & 34, Springer, Berlin, 1991.

[91] Stein C., *Efficient nonparametric testing and estimation.*, In: Proceedings of the 3rd Berkeley symposium on mathematical statistics and probability, vol. 1, University of California Press, California, Berkeley, 1956, pp. 187 – 196.

[92] Stigler S. M., *Simon Newcomb, Percy Daniel and the history of robust estimation 1885 – 1920*, Journal of the American Statistical Association **68** (1973), 872 – 879.

[93] Stigler S. M., *Do robust estimators work with real data?*, Annals of Statistics **5** (1977), 1055–1098.

[94] Stigler S. M., *Gauss and the invention of least squares*, Annals of Statistics **9** (1981), 465 – 474.

[95] Stigler S. M., *The history of statistics. The measurement of uncertainty*, Harward University Press, Massachusetts, 1986.

[96] Šidák Z., *Rectangular confidence regions for the means of multivariate normal distributions*, Journal of the American Statistical Association **62** (1967), 626 – 633.

[97] Tichavský P., *Algorithms for and geometrical characterization of solutions in the LMS and the LTS linear regression*, Computational Statistics Quarterly **8** (1991), 139 – 151.

[98] Tiku M. L. et al., *Robust inference*, M. Dekker, New York, 1987.

[99] Tukey J. W., *Explanatory data analysis*, Addison Wesley, Reading, 1977.

[100] Vajda I., *Minimum divergence principle in statistical estimation*, Statistics and Decision, Supplement Issue **1** (1984), 239-262.

[101] Vajda I., *Theory of statistical inference and information*, Reidel, Dordrecht, 1988.

[102] Velleman P. F. and Hoaglin D. C., *Efficient computing of regression diagnostics*, American Statistician **35** (1981), 234-242.

[103] Víšek J. Á., *Efficiency rate of a local deficiency of the most powerful tests in the model of contaminacy with general neighborhoods*, Kybernetika **23** (1987), 370-381.

[104] Víšek J. Á., *Estimation of contamination level in the model of contaminacy with general neighbourhoods*, Kybernetika **25** (1989), 278-297.

[105] Víšek J. Á., *A proposal of model selection: Stability of linear regression model*, In: Transactions of the 11th Prague conference on information theory, statistical decision functions and random processes, vol. B, ACADEMIA, Prague, 1991 a, pp. 451-461.

[106] Víšek J. Á., *Stability of model estimates with respect to subsamples*, Kybernetika (submitted) (1991 b).

[107] Víšek J. Á., *Testing significance of difference of models*, Kybernetika (to appear) (1991 c).

[108] Víšek J. Á., *Integral M-estimates (manuscript).*

[109] Víšek J. Á., *Adaptive estimation in linear regression model and test of symmetry of residuals*, In: Transactions of the 2nd international workshop on model-oriented data analysis, Springer, Berlin, 1991 e (to appear).

[110] Víšek J. Á., *Adaptive estimation in linear regression model, consistency and asymptotic normality*, Kybernetika (1992). (to appear)

[111] Wolke R., *Iteratively reweighted least squares. A comparison of several single step algorithms for linear models*, manuscript, Universität Halle, Halle, 1991.

[112] Wolke R. and Schwetlik H., *Iteratively reweighted least squares. Algorithms, convergence and numerical comparisons*, SIAM Journal of Scientific and Statistical Computation **9** (1988), 907-921.

[113] Wolke R. and Schwetlik H., *Rational approximations for iteratively reweighted least squares*, manuscript, Universität Halle, Halle, 1991.

[114] Yohai V. J., *High breakdown point and high efficiency robust estimates for regression*, Annals of Statistics **15** (1987), 642-656.

[115] Yohai V. J. et al., *Qualitative robustness for stochastic processes*, Annals of Statistics **15** (1987), 1293-1312.

[116] van Zwet W. R., *Van de Hulst and robust statistics. A historical note*, Statistica Neerlandica **32** (1985), 81-95.

CHARLES UNIVERSITY, DEPARTMENT OF PROBABILITY AND MATHEMATICAL STATISTICS, SOKOLOVSKÁ 83, CS-186 00 PRAHA 8, CZECHOSLOVAKIA; TEL: (+422) 231 78 62; FAX: (+422) 84 76 88; TELEX: 121 673 MFF.

CZECHOSLOVAK ACADEMY OF SCIENCES, INSTITUTE OF INFORMATION THEORY AND AUTOMATIZATION, DEPARTMENT OF STOCHASTIC INFORMATICS, POD VODÁRENSKOU VĚŽÍ 4, CS-182 08 PRAGUE 8, CZECHOSLOVAKIA, TEL: (+422) 815 20 19, FAX: (+422) 84 74 52.

E-mail: ANTOCH@CSPGUK11.BITNET and VISEK@CSPGAS11.BITNET

CONSTRUCTING PREDICTION TREES FROM DATA: THE RECPAM APPROACH

ANTONIO CIAMPI

Montreal Children's Hospital
The McGill University Research Institute

ABSTRACT. Growing trees from the data is presented as a general way of solving the prediction problem for an unknown parameter of a distribution. A tree-structured prediction model is proposed and a strategy for building such a model from the data is presented.

The strategy comprises three steps:

i) RECursive partition;
ii) Pruning;
iii) AMalgamation;

hence its acronym RECPAM. The construction is based on an information measure, the role of which is highlighted. It is shown that virtually all the available tree-growing approaches are particular cases of the general strategy.

1. INTRODUCTION

Consider the tree diagram in Figure 1 It represents a strategy for making a prediction in the context of an outpatient clinic for elderly people. Since falls in the elderly are often a cause of death, it is important to determine the risk of falling over the next year for a patient with given clinical characteristics. What the tree suggests is to ask the patient a hierarchical set of questions. The first question is: ***Do you feel dizzy sometimes?***. Every patient should be asked this question, which is shown under the ***root node***. In some sense, it is the question containing the greatest amount of information about the probability of falling. Patients answering ***yes*** to the first question, are represented

1990 *Mathematics Subject Classification*. 62H30, 62P99, 62G07, 62-02.

Key words and phrases. Prediction, classification and regression trees (CART), generalized regression trees, recursive partitioning and amalgamation (RECPAM), information increment, generalized linear model (GLIM), Cox model, outcome classification, subgroup analysis.

by the *node* at the end of the left branch issuing from the *root node*. They should be asked the next most informative question: *Do you often feel the urgency to go to the bathroom?*. Patients answering *no* to the first question (node at the end of the right branch) should be asked: *Is your gait normal?*. In this case, these three simple questions suffice, *i. e.* the prediction concerning the probability of a fall over the next year, cannot be further refined by asking more questions. This is indicated by the convention of ending the diagram with a square box. In tree jargon, square boxes are said to represent *leaves* or *terminal nodes*. Under the square boxes of the diagram, the probabilities of falling are reported. As the diagram shows, four groups are identified: low, medium, medium – high and high risk.

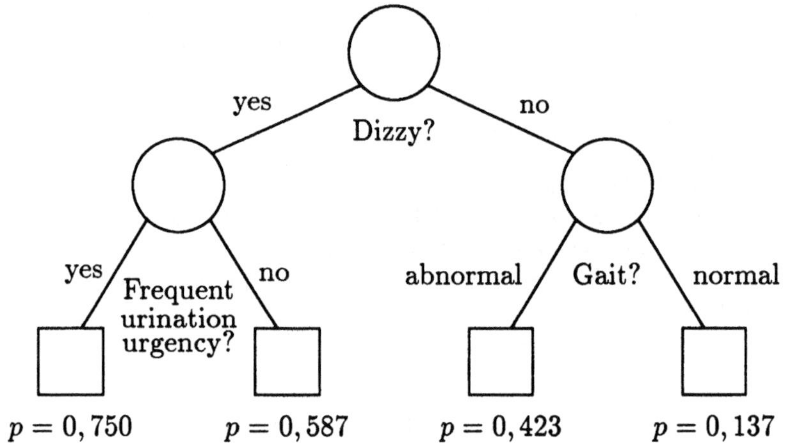

Figure 1. Prediction tree for the probability p of falling over the next year in an elderly population.

There are, of course, many other ways of summarizing the information carried by a patient's clinical status about the probability of falling. Another very common approach is to assign a ***risk score*** to a patient. In this approach, all patients are asked the same questions and a risk score is obtained, for example, summing points attributed to every *yes* answer. The risk score should then be appropriately transformed to obtain a probability of falling. In the context of this real – life problem, both a risk score and a prediction tree are available. The practicing nurses and physicians, however, seem to prefer the tree and they use it in clinical practice. In this situation, the tree's predictions happen to be more accurate than the score's.

For a statistician, the most interesting thing about the diagram in Figure 1 is that it was constructed *from data* using a computerized statistical method, which selected the three simple questions shown in the figure, out of all possible questions on 29 variables, including age and gender, variables on medical history and use of medical drugs, and variables pertaining to life-style and psycho-physical function. So was the risk score. To a statistician, risk score construction is simply *regression*, in this case, logistic regression. Tree-growing is a more recent methodology than regression. This is somewhat surprising, since tree-growing attempts to formalize what is a very natural cognitive process for decision makers. Also, an expert will often draw a tree when explaining his/her approach to problem solving. Moreover, in an approach closer in spirit to these notes, trees ("questionnaires") have been studied theoretically as vehicles of information, by Picard [23] and several authors of the French school [23].

Actually, tree-growing is not that new. It was introduced in the sixties and early seventies by the developers of AID, short for Automatic Interaction Detection, see [25] and [26]. AID predicts a continuous univariate response. A natural generalization for multivariate continuous responses was proposed in [16]. The whole methodology, however, fell into disrepute. The statistical community felt that it was not based on rigorous inference principles, and indeed, the basic issues of data driven model building were not addressed in these pioneering efforts.

Tree-growing was *rediscovered* in a biostatistical context in [5]. The prediction in that work was of a more complex nature: it was a survival curve for cancer patients and a prediction tree was constructed out of censored data. Paper [5] was a generalization of the AID procedure, but still did not address the basic issues.

If tree-growing is now taking the place it deserves in exploratory multivariate data analysis, this is due to the work of Breiman *et al.* (1984). In this monograph, a tree-building methodology for Classification and Regression Trees (CART) was described. The issues of statistical data driven modeling were clearly focused and some general solutions were proposed. Such solutions are computationally intensive and could not have been even thinkable in the sixties.

Since the seminal monograph [1] on CART, tree-growing has become popular. There are several variants of CART, such as CHAID and FACT, see [18] and [19]. These methods aim at predicting a continuous or a discrete variable. Also, various solutions to the problem of predicting a censored survival time have been proposed, see [14], [20] and [24].

It is clear from the simple example shown above, that tree-growing methods have several useful features not easily accessible by other analytic methods:
i) the ability to handle non-linearities in complex, high dimensional data sets, with both continuous and discrete predictors, and possibly with variable length records;
ii) the detection of interactions, i. e. non-homogeneities in the relationship between predictors and object of the prediction, across different subpopulations;
iii) the transparent presentation of the results of the analysis.

The purpose of these notes is to describe RECPAM, a methodology that is a generalization of CART and its variants [3] – [12]. The objective is to predict a *criterion*, on the basis of specified variables called *predictors*. By *criterion* we mean a parameter, possibly multidimensional, of a distribution of some specified variables. The distribution, however, need not be specified completely by the criterion: nuisance parameters may be needed to fully specify the distribution, and, indeed, these may be infinite dimensional.

The following are some of the unique features of RECPAM that make it the most flexible method of tree-growing to date. First, *the generality of the formulation:* this has already made it possible to treat types of data not accessible to CART and other tree-growing techniques, such as censored times and count data (Cox and Poisson regression trees), and paves the way to many more applications, some of which are currently under development. Secondly, *the distinction between global and local nuisance parameters* (see below): this permits one to treat problems such as subgroup analysis (predicting a relationship between treatment and outcome) and to develop "mixed" prediction models, *i. e.* models allowing some variables to affect the criterion linearly and some to enter the tree structure. Finally, *the amalgamation step* (see below): owing to this, the RECPAM classes are not only homogeneous in the criterion, (as the CART classes) but also distinct with respect to it, a property that is usually found in classification schemes obtained from classical clustering algorithms.

After having formulated the general prediction problem and proposed a tree – structured solution in Section 2, we outline in Section 3 the RECPAM approach. The remaining sections discuss the main realizations of the general methodology. The simplest examples are discussed in Section 4. Section 5 is devoted to a recent development that makes it possible to construct tree-structured predictors for all situations which are handled under the heading of the generalized linear model (GLIM). Section 6 discusses tree-structured survival analysis using a semi-parametric approach based on the Cox model. A discussion in Section 7 concludes the paper.

There are two appendices written by Zhiyi Lou. Appendix A gives a description of the methodology in algorithmic form, and Appendix B contains an outline of the new version of a program which implements the RECPAM methodology. The old version is extensively described in [4], [7] and [12].

2. THE GENERAL PREDICTION PROBLEM AND TREE-STRUCTURED PREDICTION

Suppose we are interested in a particular aspect of a complex phenomenon. For a statistician, a phenomenon is described by a vector random variable $\mathbf{u}$, to be observed on a number of *Observational Units* (OU's). Any interesting knowledge about the phenomenon is contained in the distribution of $\mathbf{u}, p(\mathbf{u})$, which, in principle, may vary from one OU to another. A particular aspect of the phenomenon is represented by a parametrization of $p(\mathbf{u})$ as, say, $p(\mathbf{u}; \gamma, \boldsymbol{\nu})$, where γ, the *criterion*, is a parameter representing the aspect of interest and $\boldsymbol{\nu}$, the *nuisance parameter*, represents everything else that describes the phenomenon but that is considered of secondary or no interest.

Suppose also that the phenomenon in question depends on some *external conditions*, summarized by the vector $\mathbf{z}$. This dependence can be represented by a functional dependence of γ and of $\boldsymbol{\nu}$ on $\mathbf{z}$; however, the interest is in the function $\gamma(\mathbf{z})$, with the function $\boldsymbol{\nu}(\mathbf{z})$ playing a secondary role. The general prediction problem is to estimate the function $\gamma(\mathbf{z})$ from a data matrix $\mathbf{D} = [\mathbf{U} \,|\, \mathbf{Z}]$, the rows of which, $(\mathbf{u}^{(i)} \,|\, \mathbf{z}^{(i)}), i = 1, \ldots, N$, represent measurements of $\mathbf{u}$ and $\mathbf{z}$ on N OU's, the components of $\mathbf{u}$ will be called *criterion variables* and those of $\mathbf{z}$, *predictors*. Usually, the problem is tackled in two stages:

at first, the functional form is limited to a broad class of functions, referred to as *prediction model*;

at the second stage, a particular function within that class is selected and its unknown parameters are estimated.

The first stage usually depends on a general understanding of the phenomenon and relies on the *art of modeling*. The second stage is *science* and requires direct usage of the data.

We denote by:

$$P(\mathbf{U};\boldsymbol{\gamma}(\mathbf{Z}),\boldsymbol{\nu}(\mathbf{Z})) = \prod_{i=1}^{N} p_i(\mathbf{u}^{(i)};\boldsymbol{\gamma}(\mathbf{z}^{(i)}),\boldsymbol{\nu}(\mathbf{z}^{(i)}))$$

the distribution of the sample matrix $\mathbf{D} = [\mathbf{U}\,|\,\mathbf{Z}]$. In these notes, ***estimation*** will be identified with ***optimization***. The unknown parameters will be estimated as the solution of a minimization problem:

$$(\widehat{\boldsymbol{\gamma}}(\mathbf{Z}),\widehat{\boldsymbol{\nu}}(\mathbf{Z})) = \arg\min \left\{ LF(\boldsymbol{\gamma}(\mathbf{Z}),\boldsymbol{\nu}(\mathbf{Z});\mathbf{U}) \middle| \boldsymbol{\gamma}(\mathbf{Z}),\boldsymbol{\nu}(\mathbf{Z}) \right\}$$

for a given lack-of-fit measure $LF(\boldsymbol{\gamma}(\mathbf{Z}),\boldsymbol{\nu}(\mathbf{Z});\mathbf{U})$.

The typical classical example of the general process outlined above, is *linear gaussian prediction*. In this case, in the usual regression notation, $\mathbf{u} = y$, where y is unidimensional with mean μ and variance σ^2, the criterion is $\mu = \gamma(\mathbf{z}) = \boldsymbol{\beta}'\mathbf{z}$, and the nuisance parameter is $\nu = \sigma^2$ (the usual homoscedasticity assumption implies that the nuisance parameter does not depend on $\mathbf{z}$); the sum of squares is the natural measure of lack-of-fit to be minimized. Notice that in this case the prediction model specifies both a linear relationship and the dependence of μ on *all* the predictors given in $\mathbf{z}$, which implies a linear model of a fixed number of dimensions. Then the prediction problem reduces to the estimation of the regression coefficients $\boldsymbol{\beta}$. A variant of this example is the *linear prediction model with variable selection*, in which the functional form for the criterion is $\boldsymbol{\beta}'\mathbf{z}^*$, where $\mathbf{z}^*$ denotes a subvector of $\mathbf{z}$ to be selected from the data. Now the second step includes both variable selection and estimation of the coefficients of the selected predictors.

The first stage of the approach to the prediction problem presented in these notes, consists in specifying the functional relationship $\boldsymbol{\gamma}(\mathbf{z})$ as a RECPAM tree structure, in a distribution of the general form $p(\mathbf{u};\boldsymbol{\gamma}(\mathbf{z}),\boldsymbol{\nu}(\mathbf{z}))$. Thus for the approach to be valid, it should be used when a tree-structure seems a reasonable shape for the predictor. This is often the case in complex decision making situations and, in particular, in biomedical applications, where a prediction depends on the predictors in a highly non-linear and even non-additive way. The second step consists of a search strategy for selecting a particular RECPAM tree out of the very large class of all possible structures of this kind.

3. THE RECPAM APPROACH IN GENERAL

3.1. The prediction model.

In the RECPAM tree-growing approach, we assume that there are three kinds of nuisance parameters: $\boldsymbol{\nu} = (\boldsymbol{\alpha}, \boldsymbol{\beta}, \boldsymbol{\phi})$, called *local, global*, and *residual* respectively. A local parameter is one that is known to vary very finely across the population as a function of $\mathbf{z}$. A global parameter, on the opposite end, is known not to depend on $\mathbf{z}$ and to be the same for all OU's. A residual parameter is simply one that does not enter the definition of LF, *i. e.* we can write: $LF(\boldsymbol{\gamma}(\mathbf{Z}), \boldsymbol{\nu}(\mathbf{Z}); \mathbf{U}) = LF(\boldsymbol{\gamma}(\mathbf{Z}), \boldsymbol{\alpha}(\mathbf{Z}), \boldsymbol{\beta}; \mathbf{U})$. Notice, therefore, that a criterion parameter is somewhere between a local and a global parameter. The purpose of a RECPAM analysis is to determine how local $\boldsymbol{\gamma}$ is.

The general class of RECPAM models is defined as follows. First, let us define a (*binary*) *tree*. For our purposes, this is a hierarchically structured set of (binary) questions concerning $\mathbf{z}$. The word *binary* will be omitted from now on. In what follows, a question is identified with a 0/1-valued function of the predictors, $Q(\mathbf{z})$, with $Q^{-1}(1)$ and $Q^{-1}(0)$ denoting, respectively, the subsets of the predictor space corresponding to the *yes* and *no* answers. Each question represents a node in a diagram like the one in Figure 1 A *leaf* is a terminal node and corresponds to a complete set of questions.

Next, we introduce some more notation. Let $\mathcal{P}$ denote a finite partition of the predictor space. A set of $\mathcal{P}$ will be denoted by P, and P will also be used as a running index. Let $\mathbf{I}_P(\mathbf{z})$ denote the indicator function of the set P, *i. e.* $\mathbf{I}_P(\mathbf{z}) = 1$ if $\mathbf{z} \in P$ and 0 otherwise.

We can now write down, the functional dependence of the criterion $\boldsymbol{\gamma}$ on $\mathbf{z}$ which defines a RECPAM model, as:

$$\text{(1a)} \qquad \boldsymbol{\gamma}_{\mathcal{P}}(\mathbf{z}) = \sum_{P \in \mathcal{P}} \mathbf{I}_P(\mathbf{z})\, \boldsymbol{\gamma}_P,$$

where it is assumed that the sets of $\mathcal{P}$ are unions of leaves of a tree. Thus, the leaves of a RECPAM tree represent populations which are *homogeneous* with respect to $\boldsymbol{\gamma}$, while the sets from $\mathcal{P}$ correspond to populations for which the model makes *distinct* predictions. A RECPAM tree diagram (see below) also indicates which leaves of the tree are joined to obtain distinct predictions.

As for the classical case of a linear predictor, (1a) may be interpreted as meaning that $\mathcal{P}$ is fixed and that the data determine only the parameters $\boldsymbol{\gamma}'_P$ s. Alternatively, $\mathcal{P}$ is also to be selected from the data among all possible partitions or among those of a specified subset of the set of all partitions.

To complete the specification of the RECPAM model, we need also to specify the functional form $\boldsymbol{\alpha}(\mathbf{z})$. This is done as:

$$\alpha_{\mathcal{P}}(\mathbf{z}) = \sum_{P_{Max} \in \mathcal{P}_{Max}} \mathbf{I}_{P_{Max}}(\mathbf{z})\, \alpha_{P_{Max}}, \tag{1b}$$

where $\mathcal{P}_{Max} = \{P_{Max}\}$ denotes a partition finer than $\mathcal{P}$. The subfix Max indicates that this partition, the one across which the local parameters are allowed to vary, is the maximal one considered in the construction algorithm.

3.2. Information Measures.

RECPAM is a general algorithm which selects the partitions $\mathcal{P}$ and $\mathcal{P}_{Max}$ introduced above, from a data matrix. The data are used to compute information theoretical quantities, which serve as basis for the construction. These quantities will be defined here in general. (Notice that in this section we do not use the notation $\mathcal{P}_{Max}$ since we are discussing measure of information on arbitrary partitions, denoted by $\mathcal{P}, \mathcal{P}_1, \mathcal{P}_2, \ldots$).

We start with an intuitive discussion. Suppose we consider a partition of a population into a number of subpopulations. The subpopulation label, considered as a variable, may contain some information on the criterion γ. It seems reasonable to require that a measure of this information based on the data should be non−negative (knowing the subpopulation label may be, at the worst, irrelevant to the prediction of γ), and should be larger, the greater the improvement in the prediction for γ when a distinct prediction is made for each subpopulation. Notice, however, that this requirement is reasonable only if the prediction about γ is checked on the same data used to construct the predictor. We shall define below the *observed information content of a partition* in a way that reflects this requirement, where the word *observed* emphasizes the fact that the quantity only reflects what is in the given data set.

A useful generalization is the *observed information increment of a partition with respect to a coarser one*. This definition, also given below, will reflect the following situation. Consider a population partitioned in a number of subpopulations (coarser partition), and suppose that each subpopulation can be subdivided further (finer partition). Now, we can separately predict γ, using the data, for each subset of the coarser partition and for each subset of the finer partition. It is obvious that, in the latter case, we are taking into account an amount of information concerning γ, not smaller than in the former. Thus, the information increment will be non-negative and will be larger the greater the improvement of the prediction for γ when the finer structure of the population is taken into account. The expression *observed information loss of a partition with respect to a finer one* will be used with the obvious meaning.

The important point, is that the intuitive notions of information content and information increment may be quantified and calculated from the data, given certain model assumptions. The following definition is not the most general one see [3], but it is sufficient for the purpose of these notes. The *observed information content of a partition* $\mathcal{P}$ is:

$$
\begin{aligned}
IC(\mathcal{P}\,|\,\mathbf{D}) = \inf &\left\{ LF(\gamma, \boldsymbol{\alpha}_{\mathcal{P}}(\mathbf{Z}), \boldsymbol{\beta}; \mathbf{U}); \gamma, \boldsymbol{\alpha}_P, P \in \mathcal{P} \right\} - \\
- \inf &\left\{ LF(\boldsymbol{\gamma}_{\mathcal{P}}(\mathbf{Z}), \boldsymbol{\alpha}_{\mathcal{P}}(\mathbf{Z}), \boldsymbol{\beta}; \mathbf{U}); \gamma_P, \boldsymbol{\alpha}_P, P \in \mathcal{P} \right\},
\end{aligned} \tag{2}
$$

where $\boldsymbol{\gamma}_{\mathcal{P}}$ is as in (1 a) and $\boldsymbol{\alpha}_{\mathcal{P}}$ is defined in (1 b). Thus, the observed information content is the improvement in the fit of the prediction due to allowing γ to vary across $\mathcal{P}$.

The generalization is now natural. Let $\mathcal{P}_1$ and $\mathcal{P}_2$ be two partitions with $\mathcal{P}_1$ coarser than $\mathcal{P}_2$. Then the *observed information increment of* $\mathcal{P}_2$ *with respect to* $\mathcal{P}_1$ is defined as:

$$
\Delta\mathbf{I}(\mathcal{P}_2 : \mathcal{P}_1\,|\,\mathbf{D}) = \tag{3}
$$

$$
\begin{aligned}
= \inf &\left\{ LF(\gamma_{\mathcal{P}_1}(\mathbf{Z}), \boldsymbol{\alpha}_{\mathcal{P}_2}(\mathbf{Z}), \boldsymbol{\beta}; \mathbf{U}); \gamma_P, \boldsymbol{\alpha}_{P'}, P \in \mathcal{P}_1, P' \in \mathcal{P}_2 \right\} - \\
- \inf &\left\{ LF(\gamma_{\mathcal{P}_2}(\mathbf{Z}), \boldsymbol{\alpha}_{\mathcal{P}_2}(\mathbf{Z}), \boldsymbol{\beta}; \mathbf{U}); \gamma_P, \boldsymbol{\alpha}_P, P \in \mathcal{P}_2 \right\},
\end{aligned}
$$

and, trivially, the *observed information loss* $\mathcal{P}_1$ *with respect to* $\mathcal{P}_2$ is:

$$
IL(\mathcal{P}_1 : \mathcal{P}_2\,|\,\mathbf{D}) = \Delta I(\mathcal{P}_2 : \mathcal{P}_1\,|\,\mathbf{D}). \tag{4}
$$

Notice that, if $\mathcal{P}_0$ denotes the trivial partition consisting of one set only, we have $IC(\mathcal{P}\,|\,\mathbf{D}) = \Delta I(\mathcal{P} : \mathcal{P}_0\,|\,\mathbf{D})$.

Consider now a prediction tree. By slight abuse of language, *observed information loss of a subtree* T *of the tree* T_{Max}, $IL(T : T_{Max}\,|\,\mathbf{D})$ will mean here the information loss of $\mathcal{L}_T$ with respect to $\mathcal{L}_{T_{Max}}$, where $\mathcal{L}_{T_{Max}}(\mathcal{L}_T)$ denotes the partition consisting of the leaves of $T_{Max}(T)$. Also, *observed information content of a question at a node* will mean the information content of the binary partition associated to it (*i. e.* the partition into sets comprising the OU's who answer *yes* and *no* to the question), calculated using the data for the OU's at that particular node. Similarly, $IC(T\,|\,\mathbf{D}) = IL(T : T_0\,|\,\mathbf{D})$, where T_0 denotes the trivial tree (i. e. the one – node tree consisting of the root only), will be referred to as *observed information content of the tree* T. Finally, *observed information weight at a node* will mean the information content of the partition of that node consisting of the leaves descending from it.

Before proceeding, we pause to remark that the observed information measures described so far are not ideal for measuring the performance of a predictor. A prediction should be considered as a statement based on present data but concerning *future data*. It is therefore inadequate to evaluate a prediction using present data, *i. e.* the same data used to construct the predictor. Consider, for instance, the observed information content. A better evaluation of the information content of a partition $\mathcal{P}$ about *future* data $\mathbf{D}^*$ could be based on the expected value with respect to $\mathbf{D}^*$ of:

$$(5) \qquad IC(\mathcal{P}\,|\,\mathbf{D};\mathbf{D}^*) =$$

$$= LF\left(\hat{\gamma}^1, \hat{\alpha}^1_{\mathcal{P}}(\mathbf{Z}^*), \hat{\beta}^1; \mathbf{U}^*\right) - LF\left(\hat{\gamma}^2_{\mathcal{P}}(\mathbf{Z}^*), \hat{\alpha}^2_{\mathcal{P}}(\mathbf{Z}^*), \hat{\beta}^2; \mathbf{U}^*\right),$$

where $\hat{}^1$ denotes the solution of the optimization problem:

$$\inf\left\{LF(\gamma, \alpha_{\mathcal{P}}(\mathbf{Z}), \beta; \mathbf{U}); \gamma, \alpha_P, P \in \mathcal{P}\right\},$$

based on the data set $\mathbf{D}$, and $\hat{}^2$ denotes the solution of

$$\inf\left\{LF(\gamma_{\mathcal{P}}(\mathbf{Z}), \alpha_{\mathcal{P}}(\mathbf{Z}), \beta; \mathbf{U}); \gamma_P, \alpha_P, P \in \mathcal{P}\right\}.$$

This quantity deserves the name of *predictive information content* and the analogously defined quantities $E^*[\Delta I(\mathcal{P}_2 : \mathcal{P}_1\,|\,\mathbf{D};\mathbf{D}^*)]$ and $E^*[IL\,(\mathcal{P}_1 : \mathcal{P}_2\,|\,\mathbf{D};\mathbf{D}^*)]$, that of *predictive information increment* and *loss* respectively. As remarked above, it is entirely possible that a prediction based on a finer partition might be inferior to one based on a coarser partition, when the predictions are made from data available now on *future* data. In other words, finer does not mean better when predicting the future. This is a crucial issue. It was the work of Breiman *et al.* [1] that brought such an issue into focus and proposed a reasonable way to address it using cross-validation. We will return to this question after the description of the main algorithms that constitute RECPAM.

3.3. RECPAM algorithms.

RECPAM proceeds in three steps, illustrated in Figure 2 and discussed informally below. In Appendix A, the three basic algorithms are given in pseudo-code. In what follows all information measures are supposed to be *observed*, unless explicitly stated.

Step 1: Construction of a large tree

RECPAM identifies the original population from which the data matrix has been obtained, as the root node of a tree. Starting from the root node, it searches among all the *admissible* questions of a specified class, known as the SDQ family (family of split-defining questions). The choice of the SDQ family is global and data independent. In these notes, the SDQ family is that of the *simple questions*, *i. e.* questions of the type: *Is* z_i *in A?* for a nominal predictor z_i, where A is a set of possible values; or of the type *Is* $z_i \geq a$? for an ordered predictor. Admissibility is, on the other hand, a local, data dependent condition, based on the data: a question is admissible at a node if the data at that node satisfy a certain admissibility condition (*e.g.* for a question Q to be admissible at a node, there should be no less than m individuals at that node which answer *yes* and *no* to Q). RECPAM chooses, as the first question, the one with the largest *information content*, and, as first branching, the corresponding split. The same search is performed recursively on the descendants of the root node. The process is stopped when nodes are reached with no admissible questions: this defines the leaves of the large tree, T_{Max}, *i. e.*, the tree built with no other restriction than the admissibility condition.

This step explains the first part of the acronym RECPAM: RECursive Partition.

Step 2: Pruning of the large tree and selection of the honest tree

RECPAM builds a sequence of rooted subtrees (*i. e.* subtrees containing the root node) of the large tree, beginning with T_{Max} and ending with the trivial tree (*i. e.* the subtree consisting of the root node). First, the nodes of T_{Max} are ranked in increasing order of *information weight*.

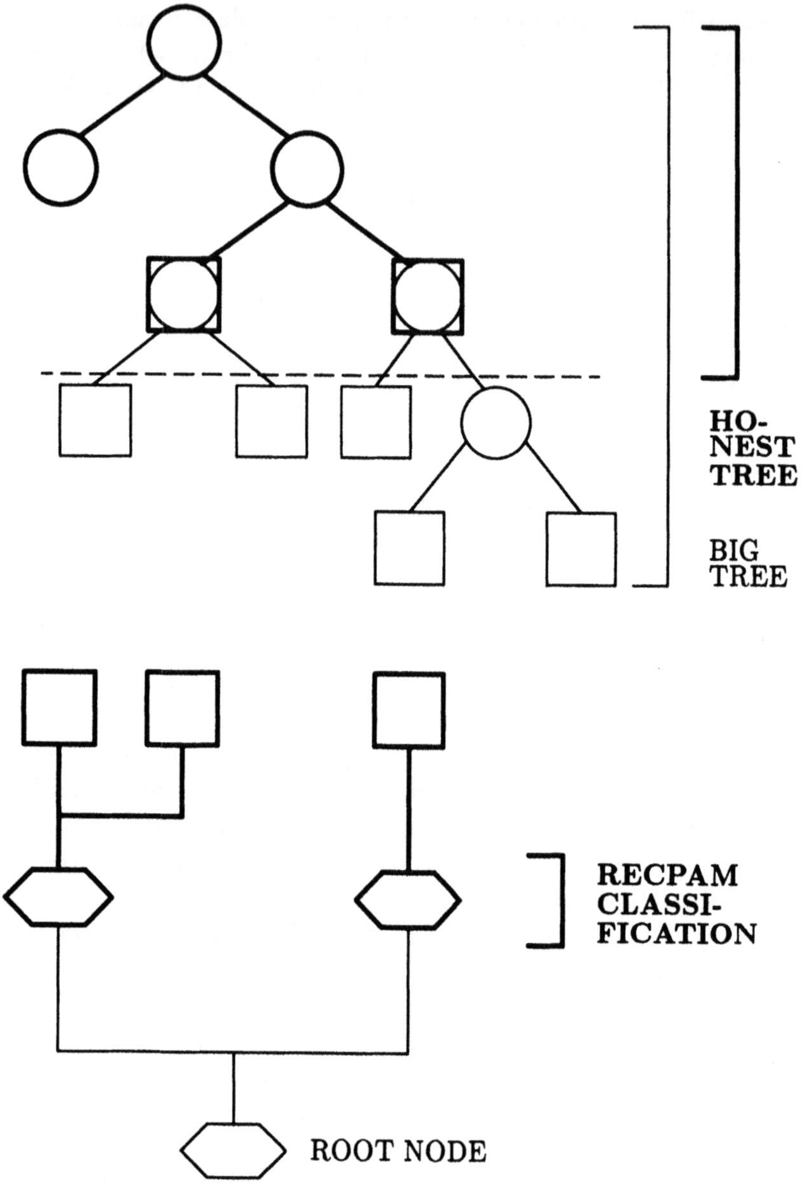

Figure 2. RECPAM scheme.

Then the sequence is constructed by going through the nodes according ing to the specified order and removing at each node the branch issuing from it. This process is known as ***pruning***.

In general, the information loss of the subtrees of the pruning sequence with respect to T_{Max} increases along the sequence. By ***honest tree*** we mean the subtree of the pruning sequence that is chosen to represent the information about γ contained in the data. In practice, the honest tree may be defined as the subtree with the largest ***predictive information content***; alternatively, it may be defined as the smallest subtree such that the information loss with respect to T_{Max} is negligible in a specified sense. Thus, a ***selection rule*** is needed to choose the honest tree out of the pruning sequence.

Step 3: Construction of the amalgamation tree and construction of the RECPAM classification

In order to obtain distinct predictions, RECPAM amalgamates the leaves of the honest tree successively, joining at each step the two subpopulations for which a minimum information loss results. This process is continued until the original population is reconstructed. The result is a structure somewhat different from the recursive partition tree, but quite similar to the hierarchies constructed by several clustering algorithms. We call this structure ***amalgamation tree***. (The term "tree" may be misleading in our context, but is standard in clustering analysis.) It is an ***ascending*** graph, in the sense that the construction proceeds from the leaves to the root. It is an ***ascending*** graph, in the sense that the construction proceeds from the leaves to the root. The partition corresponding to a step of this construction is coarser than the partitions corresponding to the preceding steps, and thus the information loss with respect to T_{Max} increases. A sequence of nested partitions results, referred to as ***amalgamation sequence***. As in step 2, a particular partition has to be chosen out of the amalgamation sequence. In this case the choice is similar to ***cutting the classification tree*** in classical cluster analysis, and the result is the RECPAM classification for predicting γ.

Depending on the particular selection rule adopted, the RECPAM classification may be defined as the partition with largest ***predictive information***, or the simplest one with negligible information loss (with respect to T_{Max}, or with respect to the honest tree).

This third step, is unique to RECPAM and explains the AM portion of the acronym.

3.4. Selecting a tree out of the pruning or the amalgamation sequence.

Both in pruning and amalgamation, we need to state a selection rule for choosing a partition out of a sequence $\{\mathcal{P}_n\}$. As discussed above, two main approaches are available. In the first approach, termed *SL approach*, one attempts to estimate how much *observed information loss*, with respect to T_{Max}, is negligible, for the given data set. When this estimate is available, one chooses the simplest (usually the smallest) partition which has negligible information loss. One situation when this approach is natural, is as follows. Suppose that, the statistics $IC(T_{Max}\,|\,\mathbf{D})$ under the *null hypothesis* of no structure (the trivial tree represents the data adequately), is distributed, at least approximately, as a known random variable S. In this case, one can apply Gabriel's theory of simultaneous tests [15] to attach to each partition of the sequence a conservative Significance Level:

$$SL(\mathcal{P}_n|\mathbf{D}) = \Pr S\,[\geq IL(\mathcal{P}_n : T_{Max}\,|\,\mathbf{D})]\,. \tag{6}$$

Now, one can consider negligible the information loss of a partition with $SL \geq \alpha$, for some a priori chosen α (in deference to tradition, one can choose in biostatistical applications $\alpha = 0.05$). It should be noted that, strictly speaking, the SL should be computed on a data set other than the one used to construct the pruning and amalgamation sequences, but in practice this is rarely done, see [3] for a discussion).

In the second approach, one considers the *observed information content* as a biased estimator of the *predictive information content*, and attempts to correct the bias by cross – validation. Cross – validation can be performed in several ways, and no universal agreement exists as to the optimal approach. The issue is discussed in [3]. Here we only emphasize that if we accept *leave – one – out cross – validation* as satisfactory, and if we have a full statistical model for the data (see, however, Section 6 for an application where the latter condition is relaxed), we can use an asymptotically equivalent computational short – cut instead of the heavy calculations of the full cross – validation approach. This is especially useful in the application of the general RECPAM approach to inherently computationally intensive situations. The short – cut leads to the simple selection rule: *choose the partition with minimum Akaike Information Criterion (AIC)*. We recall here that the Akaike Information Criterion of a statistical model given the data, is defined as:

$$AIC = \text{-2 (loglikelihood - number of estimated parameters)}.$$

The asymptotic equivalence of the minimum AIC model selection criterion to leave-one-out cross-validation was first proved by Stone [27]. Some numerical evidence in the tree-growing context can be found in [26].

3.5. RECPAM basic choices.

To summarize, a RECPAM tree construction depends on a number of user-specified choices:

Choice 1

Specification of the role of the input variables, *i. e.* which ones are the criterion and which ones are the predictors.

Choice 2

Specification of the criterion and of the measure of information increment. This often implies the specification of a statistical model for **u** conditional on a partition of the predictor space.

Choice 3

Specification of the SDQ family and of the admissibility condition at each node in Step 1.

Choice 4

Specification of the selection rule in Step 2.

Choice 5

Specification of the selection rule in Step 3.

3.6. Missing data and variable importance.

Two useful features of CART have been easily adapted to RECPAM and they will be briefly described here. We mention them here for completeness. For full details, see [1].

The first one concerns missing data. We recall that missing data can be handled in three different ways, each with its advantages and shortcomings:

i) individuals with missing measurements in any of the variables may be taken out of the data set prior to analysis;

ii) *missing* can be treated as a separate value for each variable;

iii) a missing variable can be guessed from the non-missing ones. We assume in what follows that the missing values belong to the predictors and not to the criterion variables.

Clearly, any method can handle missing data in the first way. RECPAM has an option for treating *missing* as a separate category. If this option is chosen, for each question, the category *missing* is tentatively assigned to each of the two possible answers, and definitively assigned to the one for which the information content of the question is the largest. The third option is the most interesting, since RECPAM, as CART, uses a local approach in guessing the answer to the question defining a split. At each node, the information content of all admissible questions is at first computed without the individuals with unknown answer; the question with the highest information content is selected and the corresponding split is implemented. Next, the predictors are ranked in order of their ability to predict the *best* split and their predictive ability is noted. Finally, each missing individual is assigned to the *yes* or *no* answer as follows. If the predictive ability of the first non-missing predictor surpasses a certain threshold, this predictor is used to assign the individual to the *yes* or *no* answer of the *best* split. Otherwise, the individual is assigned to the largest of the two branches (*majority rule*).

The second feature is the calculation of a *variable importance list*. This is done in order to assess the *global* contribution of each variable to the prediction. A variable which is second at each node in information content will not appear in the tree structure, yet may be extremely important as a predictor, especially since it may perform very well *everywhere*. In contrast, it is entirely possible that one variable causing a split at a certain node is predictive only at that particular node and not anywhere else. RECPAM calculates at each node of the large tree and for each variable, its information content. This is defined as the largest information content of all questions containing that particular variable. The *importance* of a variable is then computed as the sum at each node of its information content, expressed as percentage of the maximum of all such sums.

The variable importance list and the tree structure give a complementary view of the data. Variables high in the importance list and not appearing in the tree, are likely to represent global effects, that may be best taken into account in a linear regression approach. Some applications of RECPAM do permit the analyst to represent the data by models mixing a tree structure and a linear component. The variable importance list may be useful in building such models from data.

4. ELEMENTARY EXAMPLES

The simplest cases of applications of the general RECPAM approach are those in which the criterion is a parameter that completely specifies the distribution of the criterion variables, except, at most, for a *residual* parameter, not appearing in the measure of lack-of-fit.

4.1. Classification trees.

In CART, a tree such as that presented in Figure 1 is called *classification tree*, since it predicts membership to one of two or more classes - in the above example, the classes are *fallers* and *non fallers*. From the point of view of this paper, the emphasis is not on predicting class membership, but the probability of belonging to a class. In general, the criterion variable is a categorical variable y, taking nominal values labeled from 1 to k. It is assumed that, given $\mathbf{z}, y$ is multinomial, so that $\gamma(\mathbf{z})$ is the vector $\mathbf{p} = (p_1, \ldots, p_k)'$ of the probabilities of class membership, where the dependence on $\mathbf{z}$ is expressed by the model of equation (1).

The lack-of-fit measure defining the information measures of equations (3) and (4), is given by twice the negative loglikelihood under the multinomial assumption. It follows that all measures of information are likelihood ratio statistics.

The above remarks take care of choice 1 and choice 2. As for choice 3, the admissibility criterion is one of minimum multinomial cell size for each node. For instance, it is usual to require that there be at least 5 individuals in each multinomial cell for the large sample approximations to be considered valid. This is, indeed, RECPAM default option.

The selection rule in choice 4 and 5, can be chosen as the SL or the minimum AIC rule. The analyst may also perform a full cross-validation, but this approach is not yet automatically implemented in RECPAM.

4.2. Ordinary regression trees.

The application called *regression trees* in CART can be seen as an example of the general methodology presented here as follows. We assume that the criterion variable is one dimensional, $\mathbf{u} = y$, and the criterion is $\gamma = \mu = E(y)$. We search for a model of the form (1) for y given a predictor vector $\mathbf{z}$. The lack-of-fit of a model for a fixed value of γ, is measured by the sum of the squares of the deviations of the data from it. The information measures are then easily defined according to (2),(3) and (4), where LF is the sum of the squares of the deviations. Additional parameters needed to completely specify the distribution of y, such as the variance σ^2 in the case of normally distributed y, do not enter the definition of LF and are therefore residual parameters.

One of the distinguishing features of CART methodology is the use of cross-validation to correct for the overfit bias resulting from calculating the information measures on the same data set on which the parameters have been estimated. In our language, the authors of CART use the *cross-validated observed information content* of a tree, in order to obtain an estimate of the *predictive information content* as explained in Section 3.2.

This example is not implemented in RECPAM and is mentioned here only for completeness. See however the next section for implementation of a more general model for ordinary regression trees.

4.3. RECPAM for the continuous multivariate model.

In this application the criterion variable vector **u** is assumed to be multivariate normal, $\mathcal{N}(\boldsymbol{\mu}, \Sigma)$. The criterion is $\gamma = (\boldsymbol{\mu}, \Sigma)$, which implies that RECPAM classes are homogeneous and distinct with respect to the mean and the variance-covariance matrix (one or both), a more general approach than that proposed in [16].

The lack-of-fit measure is twice the negative loglikelihood of the data. As in the first example, information measures are likelihood ratio statistics. The admissibility criterion is simply minimum node size and the selection criteria are either the minimum AIC or the SL rule, with possibility of performing a full cross-validation.

An example is given in Figure 3. The tree was built from data collected during a study on the long-term sequalae of *Hemophilus Influenzae* type *b Meningitis* (HIM). The data concern 125 children who had had meningitis and were tested some years later for psycho-neurological functions [4]. It is of interest to predict the bivariate distribution of two scales, one measuring intelligence (IQ) and the other

Criteria Variables	**Description**	**Range**	**Mean**	**SE**
IQ	Wechsler intelligence scale for children - Performance I.Q. scaled scores	(65;151)	106	1,52
RS	Wide range achievements - reading standard scores	(57;142)	99	1,48

Table 1.

Predictors	**Description**	**Levels**	**Miss**
	Pre – diagnostic		
AGE	Age at admission (in month)	1 ≡ [1,12) 2 ≡ [12,27) 3 ≡ ≥ 27	0
SEX	Sex	1 ≡ Male 2 ≡ Female	0
SMPREHO	Symptomps prior to hospitalization	1 ≡ [0,1) 2 ≡ [1,2) 3 ≡ [2,4) 4 ≡ ≥ 4	1
	Diagnostic		
BGLUC	Blood glucose (ml/dl)	1 ≡ [10,95) 2 ≡ [95,135) 3 ≡ ≥ 135	13
HEMO	Hemoglobin (mg/dl)	1 ≡ [5,11) 2 ≡ [11,13) 3 ≡ ≥ 13	7
CSF_GL	Cerebral spinal fluid glucose (mg/dl)	1 ≡ [0,21) 2 ≡ [21,51) 3 ≡ ≥ 51	11
CSF_PRO	Cerebral spinal fluid protein (mg/dl)	1 ≡ [14,126) 2 ≡ [126,226) 3 ≡ ≥ 226	15

Table 1. (cont.).

Predictors	**Description**	**Levels**	**Miss**
	Post – diagnostic		
FEVDUR	In hospital fever duration (days)	1 ≡ [0,4) 2 ≡ [4,8) 3 ≡ ≥ 8	3
HOSPDUR	Length of hospitalization (days)	1 ≡ [4,10) 2 ≡ [10,14) 3 ≡ ≥ 14	0
RX	Antibiotic regimen†	1 ≡ A, C followed by A 2 ≡ A, C followed by C 3 ≡ A alone 4 ≡ C alone 5 ≡ A followed by C	1
SEQUEL	Acute–phase neurological complications	1 ≡ none 2 ≡ transient 3 ≡ persistent	0

Table 1. (cont.) † A ≡ amcipillin; C ≡ chloramphenicol.

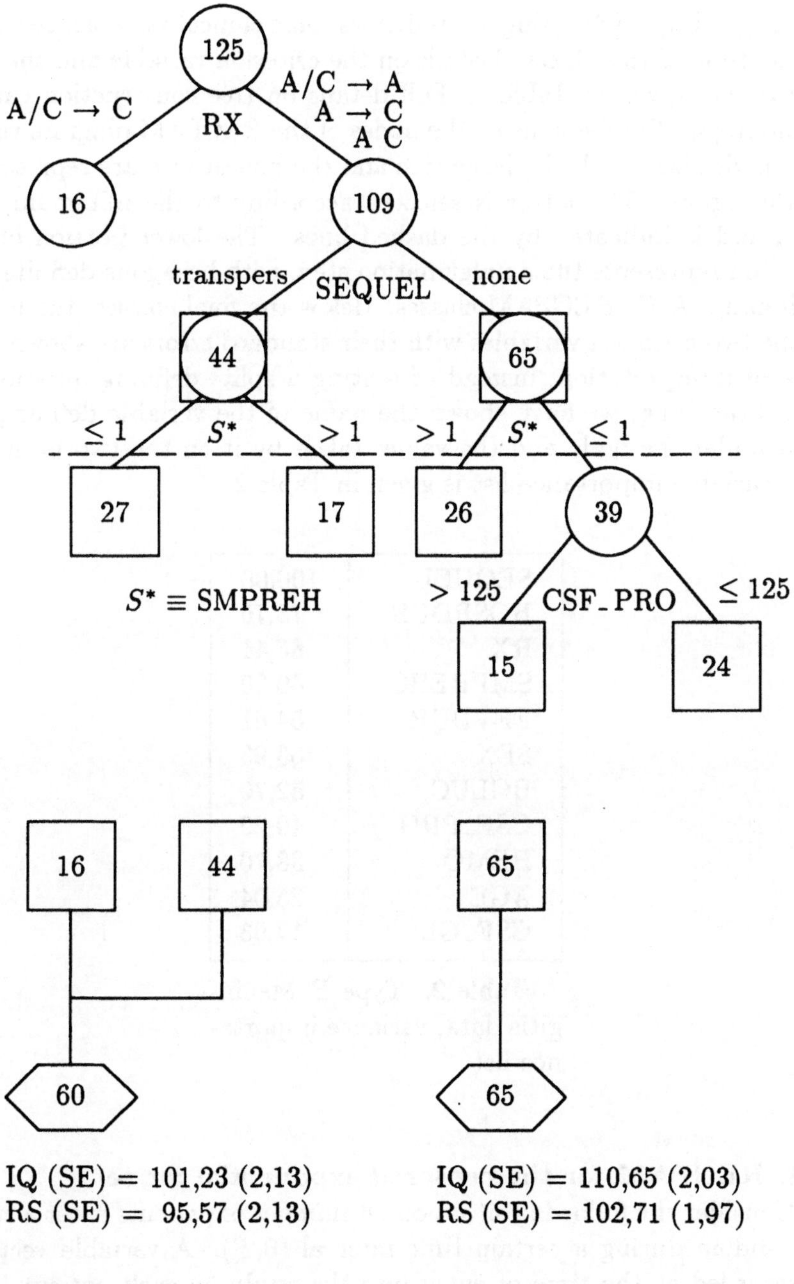

Figure 3. Outcome classification for hemophilus influenzae *Type B Meningitis*

reading ability (RS), using as predictors some clinical variables recorded at the time of the illness. Details on the criterion variable and the predictors are given in Table 1. Full details on tree construction can be found in [4]. Numbers inside the nodes of the RECPAM diagram represent node size. Both the large tree and the honest tree are represented in the figure. The latter is selected according to the minimum AIC rule, and is indicated by the dashed lines. The lower portion of the diagram represents the amalgamation step, with hexagons defining the (minimum AIC) RECPAM classes. Below the final classes, the means of the two criterion variables with their standard errors are shown. For ease of interpretation, instead of writing a split – defining question at each branching, we have shown the name of the variable defining the split under the node and the values taken by it on the two branches. The variable importance list is given in Table 2.

SEQUEL	100,00
HOSPDUR	73,76
RX	66,41
SMPREHO	60,90
FEVDUR	54,61
SEX	53,93
BGLUC	52,70
CSF_PRO	49,29
HEMO	38,70
AGE	25,04
CSF_GL	17,03

Table 2. Type B Meningitis data, variance importance list.

4.4. RECPAM for the censored exponential model.

Consider the following situation. A number of patients enter a study at random during a certain time interval $(0, S)$. A variable vector $\mathbf{z}$ is recorded at the time of entry into the study on each patient. The patients are observed during the time interval $(0, T)$, $T > S$. Some die during this time interval, others survive. Let $t^{(i)}$ denotes the length of time during which the i – th patient is observed and let $\delta^{(i)}$, the so called *censoring indicator*, be defined as taking value 1 if the patient dies at time $t^{(i)}$ after entry, and 0 if the patient is alive at that time, in which case *censoring* is said to occur. This is the typical situation of *random censoring.* It is desired to predict the survival curve of patients given $\mathbf{z}$.

The simplest approach is to assume an exponential survival curve: $S(t,\mathbf{z}) = \exp\{-h(\mathbf{z})t\}$, where $h(\mathbf{z})$ is a function of $\mathbf{z}$, not depending on t, known as *hazard*. This implies an exponential density function for the random variable: *time to death* $f(t) = h(\mathbf{z})\exp\{-h(\mathbf{z})t\}$. The loglikelihood of a data set $\{(t^{(i)}, \delta^{(i)}, \mathbf{z}^{(i)}; i = 1, \ldots, N\}$, is written as:

$$\log L = \sum_{i=1}^{N} \{-h(\mathbf{z}^{(i)})t^{(i)} + \log h(\mathbf{z}^{(i)})\delta^{(i)}\}.$$

See [17] for a detailed justification.

In the biostatistical literature, $h(\mathbf{z})$ is often assumed to be a linear function: $h(\mathbf{z}) = \boldsymbol{\beta}'\mathbf{z}$, the coefficients of which are estimated by maximum likelihood estimation. Here we assume for $h(\mathbf{z})$ the RECPAM model as in (1a):

$$h(\mathbf{z}) = \gamma_{\mathcal{P}}(\mathbf{z}) = \sum_{P \in \mathcal{P}} \mathbf{I}_P(\mathbf{z})\,\gamma_P.$$

In order to define a RECPAM algorithm of tree-construction, we specify the following basic choices:

Choice 1
The data matrix is $\mathbf{D} = [\mathbf{U}|\mathbf{Z}]$ and the criterion variables are $\mathbf{U} = \{(t^{(i)}, \delta^{(i)}); i = 1, \ldots, N\}$.

Choice 2
The criterion is the hazard. The information measures are obtained by using twice the negative loglikelihood given above as a measure of lack-of-fit.

Choice 3
The admissibility condition at each node is minimum node size and minimum number of observed deaths.

Choice 4 and 5
The minimum AIC selection rule or the SL rule may be used to choose the honest tree and RECPAM classification out of the pruning and the amalgamation sequence respectively. The SL approach is based on the large sample χ^2 approximation to the distribution of the likelihood ratio statistic for the model corresponding to the large tree with respect to the model corresponding to the trivial tree.

The example of Figure 4 was grown from data set I in [17], to which we refer the reader for a complete explanation of the available variables. Of the variables that appear in the tree, the only one needing a comment is performance status: this is a number between 0 and 100 that the physician assigns to the patient, according to certain established criteria, and that assesses the patient's position on a continuum between death (0) and perfect health (100).

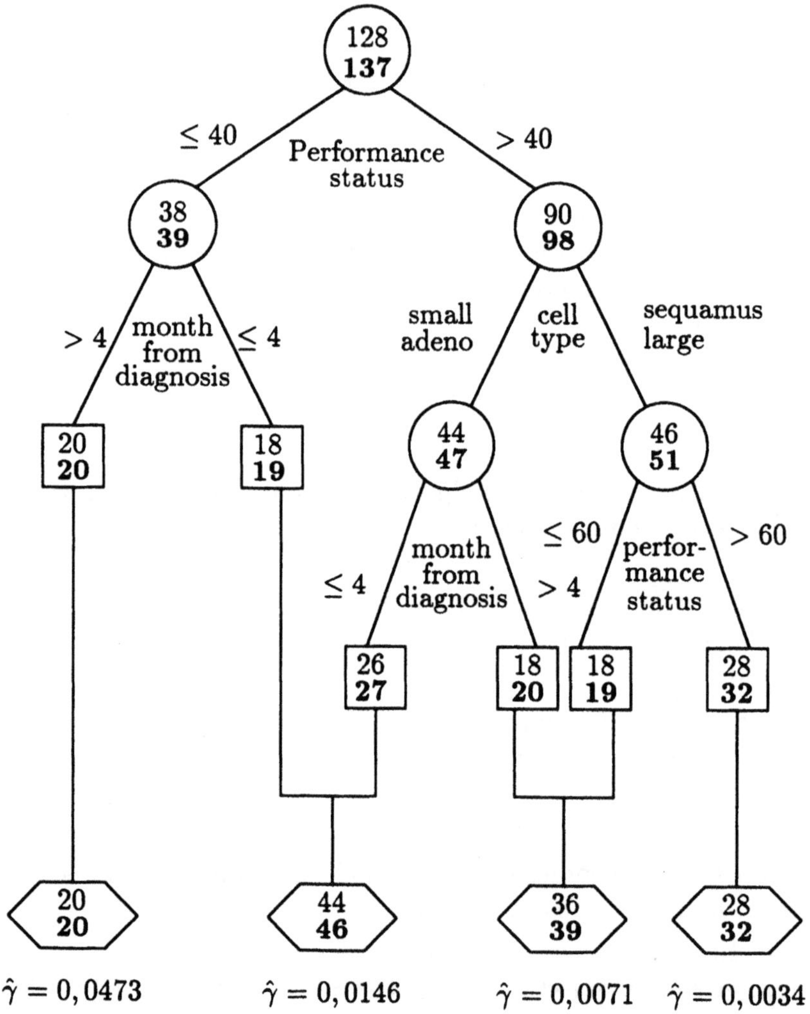

Figure 4. Outcome classification for lung cancer (outcome = number of deaths). Criterion used - hazard rate, AIC = 1432,07.

5. GLIM TREES: RECPAM FOR THE GENERALIZED LINEAR MODEL

A random variable y is said to have a distribution in the exponential family if its probability distribution is given by:

$$p(y;\theta,\phi) = \exp\left\{a(\phi)[y\theta - b(\theta)] + c(y,\phi)\right\}, \tag{7}$$

where θ and ϕ are one-dimensional parameters and a, b and c are given real valued functions. $p(y;\theta,\phi)$ is interpreted as the density if y is continuous, and as the probability function if y is discrete. In what follows, we also assume for a the form: $a(\phi) = w/\phi$.

The probability model in equation (7) is used as basis for introducing the effect of covariates of y on its distribution. It is usually assumed that the covariate vector $\mathbf{z}$ affects only the parameter θ, through a specified form $\theta(\mathbf{z})$. It is also assumed that each individual i has probability distribution:

$$p_i(y;\theta_i,\phi) = \exp\left\{(w_i/\phi)[y\theta_i - b(\theta_i)] + c_i(y,\phi)\right\}, \tag{8}$$

where w_i and the form of c_i are known. This adds flexibility in that it allows the analyst to consider situations where an OU is made up of a known number of subunits, in which case w_i, and the functional forms of θ_i and c_i are related to the size of the OU.

Great progress in applied statistics has been achieved since the introduction of covariates in (8) in a form which is linear after an appropriate scale transformation. It is not difficult to show that the expected value of y is given by the first derivative of $b(\theta) : \mu(\theta) = b'(\theta)$. In the Generalized LInear Model [21], GLIM for brevity (to be exact, the acronym GLIM refers to a popular computer package for fitting the generalized linear model, and the I stands for interactive), it is assumed that, for some given function, called ***link***, a covariate vector $\mathbf{z}$ affects (8) through:

$$link(\mu(\theta_i)) = o_i + \boldsymbol{\beta}'\mathbf{z}^{(i)} \tag{9}$$

where o_i is a unit specific given quantity known as *offset*.

Table 3. shows how various familiar regression models are specified from the general form (8), (9).

Name	**Symbol**	ϕ	$\mathbf{b}(\theta)$
Normal	$N(\mu, \sigma^2)$	σ^2	$\frac{\theta^2}{2}$
Gamma	$G(\mu, \nu)$	$\frac{1}{\nu}$	$-\log(-\theta)$
Inverse Gaussian	$IG(\mu, \sigma^2)$	σ^2	$-\sqrt{-2\theta}$
Poisson	$P(\mu)$	1	$\exp\{\theta\}$
Scaled Binomial	Bin(w,p)	1	$\log(1+\exp\{\theta\})$

Table 3. Some common distributions of the exponential family.

	$\mathbf{c}(\mathbf{y};\phi)$	$\mu(\theta)$
$N(\mu, \sigma)$	$-\frac{1}{2}\left[\frac{y^2}{\phi} + \log(2\pi\phi)\right]$	θ
$G(\mu, \nu)$	$\frac{1}{\phi}\log\left(\frac{y}{\phi}\right) - \log y - \log\Gamma\left(\frac{1}{\phi}\right)$	$-\frac{1}{\theta}$
$IG(\mu, \sigma^2)$	$-\frac{1}{2}\left\{\log(2\pi\phi y^3) + \frac{1}{\phi y}\right\}$	$\sqrt{-2\theta}$
$P(\mu)$	$-\log(y!)$	$\exp\{\theta\}$
Bin(w,p)	$\log\binom{w}{wy}$	$\frac{w\exp\{\theta\}}{1+\exp\{\theta\}}$

Table 3. Some common distributions of the exponential family (cont.).

	Canonical link (μ)	Deviance
$N(\mu, \sigma)$	μ	$\sum (y - \hat{\mu})^2$
$G(\mu, \nu)$	$-\frac{1}{\mu}$	$2\Gamma \left\{ -log\left(\frac{y}{\hat{\mu}}\right) + \frac{y - \hat{\mu}}{\hat{\mu}} \right\}$
$IG(\mu, \sigma^2)$	$-\frac{\mu^2}{2}$	$\sum \frac{(y - \hat{\mu})^2}{\hat{\mu}^2 y}$
$P(\mu)$	$\log(\mu)$	$2\sum \left\{ y \log\left(\frac{y}{\hat{\mu}}\right) - y + \hat{\mu} \right\}$
Bin(w,p)	$\log\left(\frac{\mu}{w - \mu}\right)$	$2\sum \left\{ y \log\left(\frac{y}{\hat{\mu}}\right) + (w - y) \log\left(\frac{w - y}{w - \hat{\mu}}\right) \right\}$

Table 3. Some common distributions of the exponential family (cont.).

In the GLIM approach, $\boldsymbol{\beta}$ is determined by minimizing the *deviance*. This is a very natural measure of lack-of-fit which generalizes least squares. The deviance is defined as follows. Let $\mathcal{M}$ denote any model for θ_i and let $\mathcal{M}_{Max}$ denote the *maximal model*, *i. e.*, the model which assumes a different value of θ for every OU. Let $\hat{\theta}_i$ and θ_i^* denote the maximum likelihood estimates of θ_i under $\mathcal{M}$ and $\mathcal{M}_{Max}$ respectively. Then the deviance of the model $\mathcal{M}$ from the data $\mathbf{Y}$ is:

$$\text{(10)} \qquad \text{dev}(\mathbf{Y}; \mathcal{M}) = 2\sum_i w_i \left[y_i(\hat{\theta}_i - \theta_i^*) - b(\theta_i^*) + b(\hat{\theta}_i) \right].$$

It can be shown that, if ϕ is known, then the likelihood ratio for comparing $\mathcal{M}$ and $\mathcal{M}_{Max}$ is simply $dev(\mathbf{Y}; \mathcal{M})/\phi$. It follows that maximum likelihood estimates are minimum deviance estimates. In practice, ϕ is not known. The authors of GLIM [21] suggest treating ϕ as known in estimating θ, and then estimating it *externally* as $\phi^* = D(\mathbf{Y}; \mathcal{M}^*)/(N - r)$, where $\mathcal{M}^*$ is defined as a well fitting model with r free parameters, usually a model containing all models under consideration as special cases.

In [9], the RECPAM approach was applied to GLIM. Instead of the term $\boldsymbol{\beta}'\mathbf{z}$ in (9), we assume a generalized tree structure as discussed below.

5.1. Basic Choices.

Choice 1: Criterion variables and predictors

We assume the data matrix of the form $\mathbf{D} = [\mathbf{U}|\mathbf{Z}]$, with $\mathbf{Z}$ corresponding to the predictor vector $\mathbf{z}$ and with $\mathbf{U} = [Y|W|O|\mathbf{V}^{(l)}|\mathbf{V}^{(g)}]$, corresponding to partition of the criterion variables $\mathbf{u} = (y, w, o, \mathbf{x}, \mathbf{v}^{(l)}, \mathbf{v}^{(g)})$. Of these, only the univariate y is considered random, while the others are supposed fixed (given) y is termed *outcome variable*; w and o, also univariate, are termed *weight* and *offset* respectively; the variables represented by the vectors $\mathbf{x}, \mathbf{v}^{(l)}$ and $\mathbf{v}^{(g)}$ are called *determinants*, *local confounders* and *global confounders* respectively.

With two exceptions, the terms just introduced are borrowed from [22]:
$\mathbf{Y}$ is the main object of the study, hence the *outcome* of the process under investigation;
w and o are known quantities needed to specify the distribution of $\mathbf{Y}$: their names are standard in GLIM language [21]. $\mathbf{x}$ is a *given* variable vector whose relationship to the outcome is also a principal object of the study;
$\mathbf{v}^{(l)}$ and $\mathbf{v}^{(g)}$ represent variables of secondary interest, known to affect the outcome and possibly correlated to the determinant, hence (potential) confounders: the difference between local and global confounders is that the former are assumed to effect the outcome in a way that varies very finely across the general population, so that it is desirable to correct for them as *locally* as possible, while the latter affect the outcome homogeneously across the population, hence *globally*.

Weight and offset are often absent from the data matrix, in which case it is understood that $W = 1, O = 0$.

Choice 2: Model and measure of information

As for the general GLIM approach, we take the deviance as a measure of lack of fit. This determines entirely the general RECPAM information measures (see equation (3)), provided we specify the statistical models associated to the partitions in terms of criterion and nuisance parameters.

We assume that the probabilistic model for Y is given by (8). Notice that weights w_i different from 1 occur most typically in applications in which an OU is composed of w_i subunits (*e.g.* an OU is a family and w_i the number of its member).

Given a partition of the predictor space $\mathcal{P}_1$ and a coarser partition $\mathcal{P}_2$, we assume for the link the form:

$$link\ (\mu(\theta_i)) = o_i + C(\gamma_{\mathcal{P}_1}(\mathbf{z}), \mathbf{x}) + A(\boldsymbol{\alpha}_{\mathcal{P}_2}(\mathbf{z}), \mathbf{x}, \mathbf{v}^{(l)}) + B(\boldsymbol{\beta}, \mathbf{v}^{(g)}),$$

with A, B, and C specified functions. Notice that the C term varies across the coarser partition, the A term across the finer partition and the B term does not vary at all; and that the role of the offset, as in the general GLIM approach, is to express a contribution to $link\ (\mu)$ known a priori for each OU.

With these model assumptions, the information increment measure defined by equation (3) becomes:

$$(11) \qquad \Delta I(\mathcal{P}_1 : \mathcal{P}_2 | \mathbf{D}) =$$

$$= \inf \left\{ \sum_{i=1}^{N} y_i \theta_i(\gamma_{\mathcal{P}_1}(\mathbf{Z}), \boldsymbol{\alpha}_{\mathcal{P}_2}(\mathbf{Z}), \boldsymbol{\beta}; \mathbf{U}) - \right.$$

$$\left. - \sum_{i=1}^{N} b(\theta_i(\gamma_{\mathcal{P}_1}(\mathbf{Z}), \boldsymbol{\alpha}_{\mathcal{P}_2}(\mathbf{Z}), \boldsymbol{\beta}; \mathbf{U})); \gamma_{P_1}, \boldsymbol{\alpha}_{P_2}, \boldsymbol{\beta}, P_1 \in \mathcal{P}_1, P_2 \in \mathcal{P}_2 \right\} -$$

$$- \inf \left\{ \sum_{i=1}^{N} y_i \theta_i(\gamma_{\mathcal{P}_2}(\mathbf{Z}), \boldsymbol{\alpha}_{\mathcal{P}_2}(\mathbf{Z}), \boldsymbol{\beta}; \mathbf{U}) - \right.$$

$$\left. - \sum_{i=1}^{N} b(\theta_i(\gamma_{\mathcal{P}_2}(\mathbf{Z}), \boldsymbol{\alpha}_{\mathcal{P}_2}(\mathbf{Z}), \boldsymbol{\beta}; \mathbf{U})); \gamma_{P_2}, \boldsymbol{\alpha}_{P_2}, \boldsymbol{\beta}, P_2 \in \mathcal{P}_2 \right\},$$

and from it, the remaining information measures are easily deduced.

Two general cases of (11) have been implemented so far in RECPAM:

Case 1: Outcome classification with confounders:

$$(a) \qquad A(\boldsymbol{\alpha}_{\mathcal{P}_2}(\mathbf{z}), \mathbf{x}, \mathbf{v}^{(l)}) = \boldsymbol{\alpha}_{\mathcal{P}_2}(\mathbf{z})' \mathbf{v}^{(l)},$$

$$(b) \qquad B(\boldsymbol{\beta}, \mathbf{v}^{(g)}) = \boldsymbol{\beta}' \mathbf{v}^{(g)},$$

(c) $$C(\gamma_{\mathcal{P}_1}(\mathbf{z}), \mathbf{x}) = \gamma_{\mathcal{P}_1}(\mathbf{z}).$$

Here we wish to study the dependence of the expected value of $\mathbf{y}$ on the predictors, in the presence of both local and global confounders.

Case 2: Subgroup analysis with confounders:

(a) $$A(\alpha_{\mathcal{P}_2}(\mathbf{z}), \mathbf{x}, \mathbf{v}^{(l)}) = \alpha^{(1)}_{\mathcal{P}_2}(\mathbf{z}) + \alpha^{(2)}_{\mathcal{P}_2}(\mathbf{z})'\mathbf{v}^{(l)},$$

(b) $$B(\beta, \mathbf{v}^{(g)}) = \beta'\mathbf{v}^{(g)},$$

(c) $$C(\gamma_{\mathcal{P}_1}(\mathbf{z}), \mathbf{x}) = \gamma_{\mathcal{P}_1}(\mathbf{z})'\mathbf{x}.$$

The goal is, in this case, to study the effect of the predictors as possible modifiers of the relationship between y and $\mathbf{x}$, in the presence of both local and global confounders.

Choice 3: Admissibility condition

Let $Q_i, i = 1, 2$, denote the sets of the partition generated by the question Q, and let $\mathbf{D}_{Q_i}$ denotes the portion of the data matrix corresponding to Q_i. The following admissibility conditions have been implemented in RECPAM:

$$\mathbf{1}'\mathbf{W}_{Q_i} \geq m_1 \text{and} \mathbf{1}'\mathbf{Y}_{Q_i} \geq m_2.$$

They can be interpreted as restrictions on minimum leaf size and minimum count size for the leaves of the large tree.

Choice 3 and 4: selection rules for the pruning and amalgamation sequences

Both minimum AIC rule and SL rule have been implemented. As discussed above, this requires knowing the loglikelihood of the statistical model associated to the trees and partitions which appear in the pruning and amalgamation sequences. Now, twice the negative of the loglikelihood is just the deviance divided by ϕ. If ϕ is to be estimated from the data, this is done by dividing the deviance for the model corresponding to the large tree by $N - k_{Max}$, where k_{Max} denotes the number of parameters associated to the large tree.

5.2. Applications: An example.

By appropriately specifying the functions a, b and c in the definition of the exponential family and the link function, a great variety of common statistical models can be included in the RECPAM approach. For each of them, both outcome classification and subgroup analysis can be developed according to Case (a) and Case (b) defined above. Tree-structured normal, logistic and Poisson regressions and tree-structured regression models for censored survival data have been discussed in [9]. Notice that the case of the outcome classification for the normal model, yields a tree construction algorithm analogous to CART's regression tree approach. We will discuss here the case of Poisson regression through an example using real data.

Consider the data in Appendix VII of [2]. They concern nasal sinus cancer mortality in a cohort of Welsh nickel refinery workers. The OU's are groups of workers, homogeneous with respect to age at first employment (AFE), year of first employment (YFE), time of first employment (TFE) and exposure level (EXP). For each OU, we have the number of deaths n and person-years (PY). The outcome is n. Table 4 gives the categories of the variables AFE, YFE, TFE and EXP which are given in the data.

AFE	YFE	EXP	TFE
$1 \equiv\ \leq 20{,}0$	$1 \equiv$ 1902-1910	$1 \equiv 0$	$1 \equiv$ 0-19,9
$2 \equiv$ 20,0-27,4	$2 \equiv$ 1910-1914	$2 \equiv$ 0,5-4	$2 \equiv$ 20-29,9
$3 \equiv$ 27,5-34,9	$3 \equiv$ 1915-1919	$3 \equiv$ 4,5-8	$3 \equiv$ 30-39,9
$4 \equiv$ 35,0-54,4	$4 \equiv$ 1920-1924	$4 \equiv$ 8,5-12	$4 \equiv$ 40-49,9

Table 4. Categories of classification variables for nasal cancer analyses.

A first analytic problem is to coherently organize the useful information concerning the risk of developing nasal cancer, which is contained in the variables AFE, YFE, EXP and TFE (the predictors). This is an outcome classification problem without confounders. The criterion variables are the outcome n and PY. It is natural to assume n Poisson with mean $PY\,e^{\gamma}$, where γ is the criterion. Then PY plays the role of offset, as it would in a linear regression model [21]: thus $o = PY$, and the weight can be set to 1. See Table 3 for the specification of the Poisson model in GLIM.

In Figure 5 we give the outcome classification tree obtained from the data. The admissibility condition is specified as a minimum number of OU's and of nasal cancers for the leaves of the large tree. The minimum AIC rule was used for pruning and amalgamation.

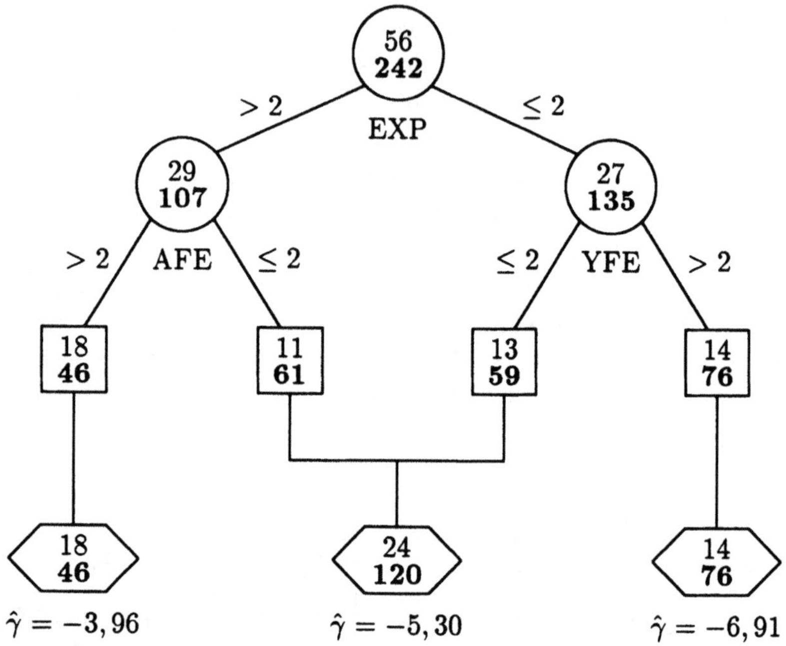

Figure 5.Outcome classification for nasal sinus cancer (outcome = death). Criterion: $\hat{\gamma} = ln\ [\ Mean/PY]$.

Since exposure plays a crucial role, it is interesting to see whether its relationship to outcome depends on the other variables. This leads to subgroup analysis. The role of the variables is now different, in that the criterion variables are n (outcome), PY (offset) and EXP (determinant). There are, again, no confounders. We assume now that n is Poisson with mean $PY\ \exp\{\alpha + \gamma\ EXP\}$, where α is a local parameter and γ the criterion. The predictors are AFE, YFE and TFE. The reason why α is defined as local is that the predictors are likely to influence the outcome directly, as well as through their effect on the relationship between exposure and outcome.

The tree of Figure 6 gives the result of the subgroup analysis.

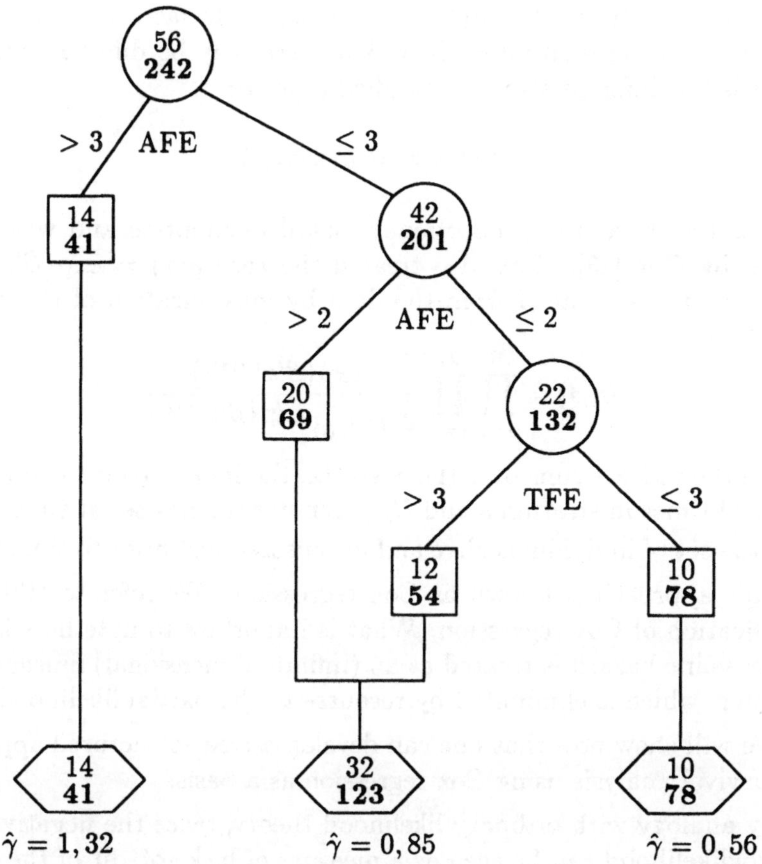

Figure 6. Subgroup analysis for nasal sinus cancer (determinant = exposure). Criterion: coefficient of exposure.

6. TREE-STRUCTURED SURVIVAL ANALYSIS WITH THE COX MODEL

In Section 4.5, we have discussed the construction of prognostic classification trees when the outcome of interest is a censored survival time that can be assumed exponential. We now return to survival analysis with censoring under the more general assumption that the distribution of the survival time is unknown, but that the prognostic factors act on survival by altering the baseline hazard $h_0(t)$ in a simple way (the hazard is simply the logarithmic derivative of the cumulative distribution function of the survival time, and the baseline hazard is the hazard corresponding to $\mathbf{z} = 0) : h(t; \mathbf{z}) = g(\mathbf{z})\ h_0(t)$. In fact, we consider the even more general situation in which there may be different baseline hazards for different strata s specified a priori:

$$h(t; s, \mathbf{z}) = g(\mathbf{z}) h_s(t)$$

This is known as the proportional hazard assumption and was introduced by Cox [13]. Cox also treated the case $g(\mathbf{z}) = \exp\{\boldsymbol{\beta}'\mathbf{z})$ and proposed to estimate $\boldsymbol{\beta}$ from the data by maximization of the *partial likelihood*:

$$L(\boldsymbol{\beta}) = \prod_{s=1}^{S} \prod_{i=1}^{d_s} \frac{\exp\{\boldsymbol{\beta}'\mathbf{z}^{(s(i))}\}}{\sum_{l \in R_{s(i)}} \exp\{\boldsymbol{\beta}'\mathbf{z}^{(s(l))}\}},$$

where the index s runs over the S strata, the index i over the observed survival times in stratum s and $R_{s(i)}$ denotes the risk set at time $t^{(s(i))}$, *i. e.* the set of individuals alive and uncensored just prior to time $t^{(s(i))}$.

This approach is known as Cox regression. We refer to [17] for a justification of Cox regression. What is important to note here is that the baseline hazard is treated as an (infinite dimensional) nuisance parameter, which is eliminated by recourse to the partial likelihood.

We will show now that one can develop a tree – structured approach to survival analysis using Cox regression as a basis.

By analogy with ordinary likelihood theory, twice the negative partial loglikelihood can be seen as a measure of lack – of – fit of the linear portion of the model, given the proportional hazard assumption. The RECPAM developments are based on this analogy.

As in the previous section, we will consider two analytical problems: prognostic classification and subgroup analysis. There will be neither local nor global parameters.

6.1. Tree – structured prognostic classification.

Consider data obtained from recording on N individuals the variables: $(t, \delta, s, \mathbf{z})$, where t and δ are as in Section 4.4, s is a stratum indicator (for instance, a stratum may be a hospital in which a patient is treated), and $\mathbf{z}$ denotes a vector of (potential) predictors of survival. The problem is to summarize the prognostic information contained in $\mathbf{z}$ by means of a tree – structured classification, while correcting for the stratum variable. The RECPAM approach is developed as follows.

Choice 1

The criterion variables are (t, δ, s) and the predictors are $\mathbf{z}$.

Choice 2

The criterion is defined by the equation:

$$h(t; s, \mathbf{z}) = h_s(t) \exp\{\gamma(\mathbf{z})\}.$$

Thus, the criterion is the logarithm of the relative hazard with respect to the (stratified) baseline population. Notice that, since RECPAM aims at finding a model of the form (1), the baseline has to be defined as a specific element of the partition $\mathcal{P}$. The convention implemented here is that the baseline is the set of the partition corresponding to the best prognosis.

Using twice the negative partial loglikelihood as measure of fit, it is easy to see that the natural measures of information are (partial) likelihood ratio statistics comparing models corresponding to different trees and/or partitions.

Choice 3

The admissibility condition is a restriction on the minimum size of a leaf, and/or on the minimum number of observed deaths in a leaf.

Choice 4 and 5

In analogy to ordinary likelihood theory, the minimum AIC rule and the SL approach are implemented by simply treating the partial loglikelihood as an ordinary likelihood.

Criteria Variables	**Levels**
Survival	Days
Status	0 ≡ Censored 1 ≡ Dead
Institution	1 ≡ Inst1 2 ≡ Inst2 3 ≡ Inst3 4 ≡ Inst4 5 ≡ Inst5 6 ≡ Inst6
Treatment	1 ≡ Standard 2 ≡ Test

Table 5. Variable description for oropharynx data.

Predictor Variables	Levels
Age	Range ≡ [20–90]; Mean = 60,44; SE=11,21
Sex	1 ≡ Male 2 ≡ Female
Condition	1 ≡ No disability 2 ≡ Restricted work 3 ≡ Requires assistance with self care 4 ≡ Bed confined
Grade	1 ≡ Well differentiated 2 ≡ Moderately differentiated 3 ≡ Poorly differentiated
Site	1 ≡ Faucial arch 2 ≡ Tonsillar fossa 3 ≡ Posterior pillar 4 ≡ Pharyngeal tongue 5 ≡ Posterior wall
T staging	1 ≡ Primary tumor measuring 2 cm or less in largest diameter 2 ≡ Primary tumor measuring 2 cm to 4 cm in largest diameter, minimal infiltration in depth 3 ≡ Primary tumor measuring more 4 cm 4 ≡ Massive invazive tumor
N staging	0 ≡ No clinical evidence of mode metastases 1 ≡ Single positive node, 3 cm or less in diameter, not fixed 2 ≡ Single positive node, more than 3 cm in diameter, not fixed 3 ≡ Multiple positive nodes or fixed nodes

Table 5. Variable description for oropharynx data (cont.).

Figure 7 gives an example of prognostic stratification obtained from the data set II of reference [17]. To facilitate the interpretation of the figure, the variables used in the analysis are defined in Table 5.

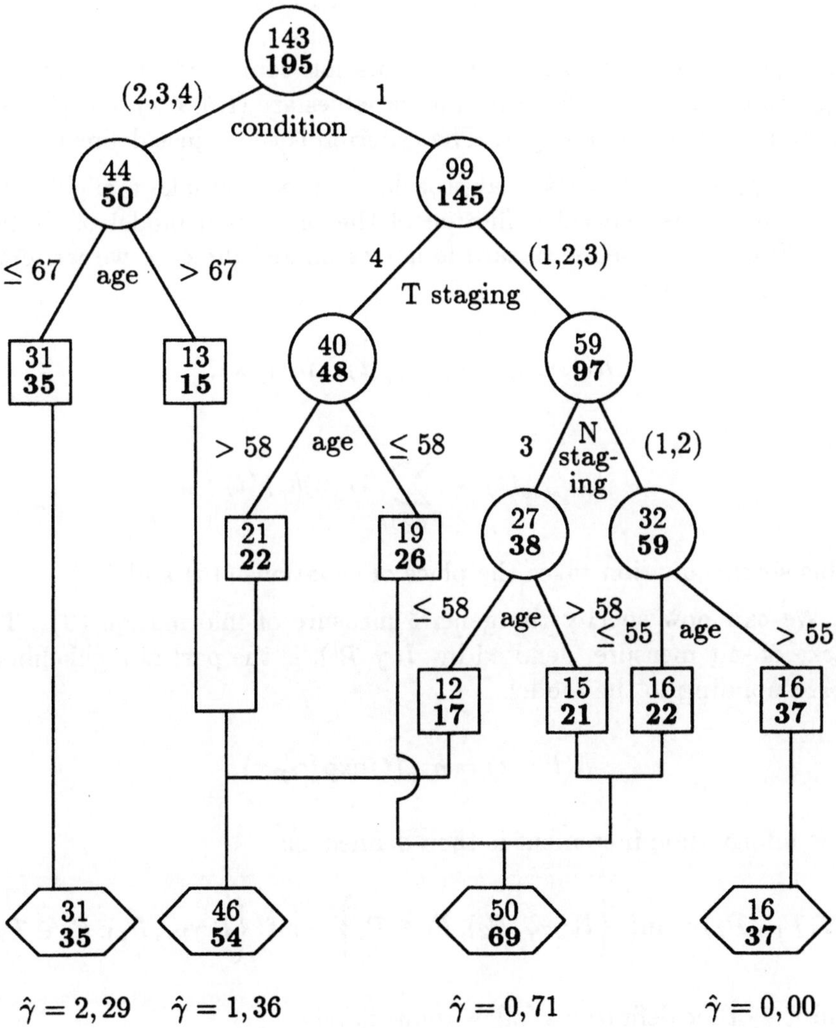

Figure 7. Outcome classification for cancer of the oropharynx (outcome = number of deaths). Criterion used: log[*Relative Hazard*. AIC = 785,51.

6.2. Subgroup analysis.

Consider data from variables of the form: $(t, \delta, s, \mathbf{x}, \mathbf{z})$, where t, δ, and s are as above, $\mathbf{x}$ is a vector of *determinants* of survival, and $\mathbf{z}$ a vector of potential *modifiers* of the relationship between survival and $\mathbf{x}$. We assume that this relationship is well described by a Cox model:

$$h(t; s, \mathbf{z}) = h_s(t) \exp\{\gamma' \mathbf{x}\}.$$

The goal is to summarize the information contained in $\mathbf{z}$ about the regression vector γ. The criterion variables are $(t, \delta, s, \mathbf{x})$ and the predictors are represented by $\mathbf{z}$. The criterion is the γ just defined.

To apply the RECPAM approach, we need to make a slight modification to the general definition of the prediction model of Section 3.1. Since the baseline hazard is not estimated directly, we write the prediction model as:

$$h(t; s, \mathbf{z}) = h_{s\mathcal{P}_{Max}}(t) \exp\{\gamma'_{\mathcal{P}} \mathbf{x}\},$$

where

$$h_{s\mathcal{P}_{Max}}(t) = \sum_{P \in \mathcal{P}_M} \mathbf{I}_P(\mathbf{z}) h_{sP}(t).$$

This single equation takes the place of equations $(1a)$ and $(1b)$.

We can now specify the general measure of information (3). The lack-of-fit measure, denoted by $l(\gamma, \mathcal{P}')$, is the partial loglikelihood corresponding to the model:

$$h(t; s, \mathbf{z}) = h_{s\mathcal{P}}(t) \exp\{\gamma'_{\mathcal{P}} \mathbf{x}\}.$$

The information increment is then defined as:

$$\Delta I(\mathcal{P}_1 : \mathcal{P}_2) = \inf \left\{ l(\gamma_{\mathcal{P}_1}, \mathcal{P}_2); P_1 \in \mathcal{P}_1 \right\} - \inf \left\{ l(\gamma_{\mathcal{P}_2}, \mathcal{P}_2); P_2 \in \mathcal{P}_2 \right\}$$

and the other definitions follow immediately.

The other basic choices which define a RECPAM algorithm are as for the prognostic classification.

In Figure 8 we give the subgroup analysis for treatment effect for data set II of reference [17]. As it can be seen, the data suggest that sex is the only modifier of the treatment effect: for women, the two treatments have practically the same effect, while for men the treatment coded as 1 (see Table 5) is definitely superior to the other one (coded as 2).

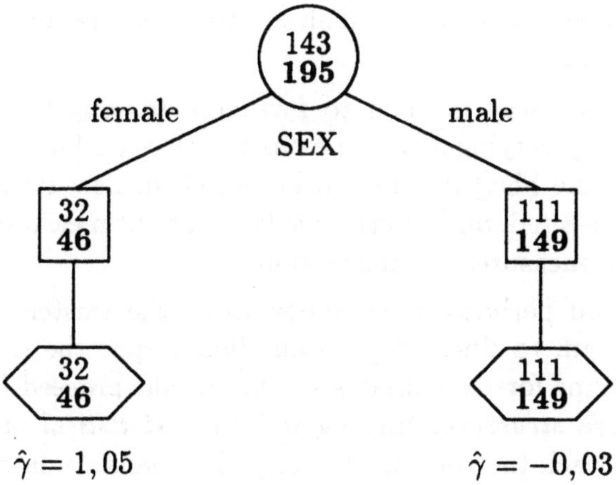

Figure 8. Subgroup analysis for cancer of the oropharynx (determination=treatment). AIC=471,576. Criterion: $\hat{\gamma}$ – coefficient of treatment.

7. CONCLUDING REMARKS

These notes are only an introduction to an area of methodological research that is attracting an increasing amount of interest. We have not attempted an exhaustive review of the field. Instead, we have shown that tree–growing can be reduced to a few basic principles and choices.

Tree construction has been presented here as a strategy of *information seeking* from the data. This is characterized by the assumption of a certain model corresponding to a natural *cognitive structure*. Within this basic choice, various approaches have to be devised to assure that the information sought is *generalizable*.

The generality of the approach and the wealth of available examples of data analysis, should invite the reader to think of tree–growing as an alternative (or complementary) paradigm to (generalized) linear regression. Practically everything that can be done with linear regression, can also be done with tree–growing. The question to ask in a given situation is then: which of the two paradigms is the most appropriate one? The ideas associated to cross–validation, only roughly sketched here, owing to space limitations, offer the basic tools to answer such a question.

We feel that tree–growing constitutes extraordinarily fertile ground for future research. First, there are many detailed questions to answer concerning cross – validation and alternative pruning and amalgamation strategies (see 5).

Secondly, the way is open to further extending the scope of the methodology by varying any of the basic choices outlined in these notes: for example, the SDQ family can be enriched, new statistical models can be implemented, and alternative lack – of – fit measures can be used to define new measures of information.

Thirdly, and perhaps most importantly, the existence of an alternative paradigm to linear regression should spur the development of other paradigms for data analysis. One would proceed by selecting a basic *cognitive structure*, turning it into a *statistical model*, and attempting to teach the computer to organize available data according to that structure, in a way that is *generalizable* to other similar data. As computerized information becomes increasingly available at reasonable cost, this approach appears both natural and far reaching. It may well provide a solid framework for the much touted convergence of Statistics and Artificial Intelligence.

APPENDIX A.

Tree Building.

P and P' are Observation Units
IC is Information Content
VI is Variable Importance

↓
Enter data matrix
↓
Set the stopping rule (admissibility criterion)
↓
root = build _ tree $(1, P)$
↓
Generate new node
↓
Check the stopping rule
↓

↓	↓
If it is satisfied	*If it is not satisfied*
↓	↓
node → split _ variable = 0	node → split _ variable = predictor with maximum IC in all predictors

↓	↓
node → split _ value = 0	node → split _ value = split point with maximum IC for node → split _ variable
↓	↓
node → left = NULL	split $(1, P)$ into two groups: $(1, P')$ and $(P'+1, P)$
↓	↓
node → right = NULL	If predictor is missing, do surrogate
↓	↓
return	node → left = build _ tree $(1, P')$
	↓
	node → right = build _ tree $(P'+1, P)$
	↓

↓

Computing VI for all predictors

↓

Output large tree

↓

Tree Pruning.

IW is information weight
IL is information loss
SL is significance level

↓	↓
Enter data matrix	Enter tree structure
↓	↓

↓

Pruning rule stopping rule (honest tree criterion)

↓

Read tree

↓

Calculate IW for each internal node

↓

Sort IW in increasing order

↓

⟶ +

↓

According to IW order, pruning one internal node

↓

Estimate all parameters and calculate AIC, IL, SL

↓

Check pruning stopping rule

↓

+ ⟵ If it is not satisfied

↓

If it is satisfied

↓

Output pruned tree

Tree Amalgamation.

IL is information loss

SL is significance level

↓ ↓

Enter data matrix   Enter tree structure

↓ ↓

↓

Set Amalgamation stopping rule (RECPAM classification criterion)

↓

Read tree

↓

⟶ *

↓

Consider all possible combination of pairs of two leaf nodes which have different parents

↓

Choose the combinations with minimum IL by comparing the IL for each combination

↓

Estimate all parameters and calculate AIC, IL, SL

↓

Check amalgamation stopping rule

↓

* ⟵ If it is not satisfied

↓

If it is satisfied

↓

Output amalgamated tree

↓

APPENDIX B

Implementation of RECPAM.

The RECPAM software package implements the general RECPAM approach to tree-growing, and provides some utilities to support its implementation, such as data preparation and tree output. It works under a menu driven interactive environment for data analysis. It is easy to use and flexible, allowing the user to make selections. So far, we have developed four statistical models, i. e., *Multinomial, exponential, GLIM* and *COX*. A general and consistent algorithm for developing other statistical models is available. An old version also contains a multivariate model.

The implementation of the RECPAM approach is the core of the RECPAM package. It is organized into three independent functions: tree building, tree pruning and leaf amalgamation.

Tree Building

Allows the user to:

- recursively construct a partition with maximal information content about parameters to be predicted (criterion);
- process the missing data using surrogate method, if there are missing predictors;
- give the variable importance list of predictors.

The user specifies three parameters:

- name of the data file to be analyzed which has been prepared by a data preparation utility;
- name of the large tree file which will be returned by the program;
- the stopping rule which includes two admissibility criteria: observation size of leaf node and/or event size of leaf node.

The program returns two files:

- large tree file which presents the construction of a large tree;
- default result file in *larger.res* which includes the variable importance list of all predictors and surrogate coefficients of all predictors and surrogate coefficients of all predictors for each split.

Tree Pruning

Allows the user to:

- calculate the information weight of each internal node and rank them in order of increasing information weight in order of increasing information weight in order to define the pruning order;

- estimate coefficients of all parameters and calculate AIC, information loss and significance level for each pruning step;
- select the honest tree by choosing the stopping rule.

The user specifies four parameters:
- data file for analysis;
- corresponding large tree file name;
- name of the pruned tree which will be returned by the program;
- stopping rule for honest tree: AIC, Significance level or User.

The program returns two files:
- pruned tree file which presents the construction of a honest tree;
- default result file in *prune.res* which includes estimation and standard error of the criterion for each leaf, AIC, information loss and significance level for each pruning step.

Leaf Amalgamation

Allows the user to:
- consider all possible combinations of pairs of leaf nodes which have different parents and choose the combinations with minimum information loss by comparing the information loss for each combination (this is repeated recursively until a stopping rule is satisfied);
- estimate coefficients of all parameters and calculate AIC, information loss and significance level for each pruning step;
- select the RECPAM partition by choosing the stopping rule.

The user specifies four parameters:
- data file for analysis;
- corresponding large tree file name or pruned tree file name;
- name of the amalgamation file which will be returned by the program;
- stopping rule for RECPAM partition: Minimum AIC, Significance level or user.

The program returns two files:
- amalgamation tree file which presents the combination of leaves of a tree — it is a partition rather than a tree structure;
- default result file in *amalg.res* which includes estimations and standard errors of criterion for each leaf, AIC, information loss and significance level for each amalgamation step.

The supporting utilities of RECPAM include data preparation, data regression, tree entering, tree printing and tree drawing. They provide user interface between data and the RECPAM approach, and increase the flexibility of data analysis.

Data Preparation

Allows the user to:

- select one of the statistical models;
- specify the size, names and types for the criterion and predictors;
- validate the data file.

User specifies two parameters:

- name of the raw data file (an ASCII data file);
- name of the prepared data file which will be returned by the program.

The program returns one file:

- prepared data file which may be used for any data analysis.

Tree Printing

Translates the tree – structure to text representation using the pre – order visit method.

The user specifies two parameters:

- name of big tree file or pruned tree file which contains a structure;
- name of tree text representation file which will be returned by the program.

The program returns one file:

- tree – structured text presentation file.

Tree Drawing

Translates the tree-structure to graphic representation, but only for small trees, due to screen and printer limitations.

The user specifies two parameters:

- name of large tree file, *pruned tree* file or amalgamation file;
- name tree graphic representation file.

The program returns one file:

- tree graphic representation file.

Enter Tree

Allows user to create a tree structure or use an existing tree structure for pruning, amalgamation or computing AIC estimations.

The user specifies three parameters:

- data file for analysis;
- tree structure file name (if the user defines the tree structure, enter user as tree structure file);

- name of generating tree structure file which will be returned by the program.

The program returns one file:

- tree structure file.

Data Regression

Does regression analysis and returns regression coefficient estimation and loglikelihood

The user specifies one parameter:

- name of data file for analysis.

The program returns one file:

- default result file in regression which includes estimations and standard errors of all parameters and loglikelihood.

ACKNOWLEDGEMENTS

The author's research on regression trees is supported by a grant from the Medical Research Council of Canada. He gratefully acknowledges the enthusiastic and careful assistance of Lisa Hendricks and Abdissa Negassa in producing this manuscript.

REFERENCES

[1] Breiman L., Friedman J. H., Olshen R. A. and Stone C. J., *Classification and regression trees*, Wadsworth, Belmont CA, 1984.

[2] Breslow N. E. and Day N. E., *Statistical methods in cancer research*, IARC Scientific Publications (Heseltine E., ed.), vol. 2 (82), Oxford University Press, Oxford, 1987.

[3] Ciampi A., *Generalized regression trees*, Computational Statistics & Data Analysis **12** (1991), 57 – 78.

[4] Ciampi A., du Berger R., Taylor H. G. and Thiffault J., *RECPAM: A computer program for recursive partition and amalgamation for censored survival data and other situations frequently occurring in biostatistics. III. Classification according to a multivariate construct. Application to data on Haemophilus influenzae type b meningitis*, Computer Methods and Programs in Biomedicine **36** (1991), 51 – 61.

[5] Ciampi A., Bush R. S., Gospodarowicz M. and Till J. E., *An approach to classifying prognostic factors related to survival experience for non – Hodgkins lymphoma patients*, Cancer **47** (1981), 621 – 627.

[6] Ciampi A., Chang C. H., Hogg S. A. and McKinney A., *Recursive partition: A versatile method for exploratory data analysis in biostatistics*, Biostatistics (MacNeil I. B. and Umphrey G. J., eds.), 1987, pp. 23 – 50.

[7] Ciampi A., Hogg S. A., McKinney S. and Thiffault J., *RECPAM: A computer program for recursive partition and amalgamation for censored survival data*

and other situations frequently occurring in biostatistics. I. Methods and program features, Computer Methods and Programs in Biomedicine **26** (1988), 239 – 256.

[8] Ciampi A., Lawless J. F., McKinney S. M. and Singhal K., *Regression and recursive partition strategies in the analysis of medical survival data*, Journal of Clinical Epidemiology **41** (1988), 737 – 748.

[9] Ciampi A., Lou Z., Lin Qian and Negassa A., *Recursive partition and amalgamation with the exponential family: theory and applications*, Applied Stochastic Models and Data Analysis **7** (1991), 121 – 137.

[10] Ciampi A. and Thiffault J., *Recursive partition and amalgamation (RECPAM) for censored survival data: criteria for tree selection*, Statistical Software Newsletter **14** (1988), 78 – 81.

[11] Ciampi A. and Thiffault J., *Pruning regression trees for censored survival data: the RECPAM approach*, Communications in Statistics – Theory and Methods **18 (9)** (1989), 3373 – 3388.

[12] Ciampi A., Thiffault J. and Sagman U., *RECPAM: A computer program for recursive partition and amalgamation for censored survival data and other situations frequently occurring in biostatistics. II. Applications to data on small cell carcinoma of the lung (SCCL)*, Computer Methods and Programs in Biomedicine **30** (1989), 283 – 296.

[13] Cox D. R., *Regression models and life tables*, Journal of the Royal Statistics Society **B 34** (1972), 187 – 208.

[14] Davis R. B. and Anderson J. R., *Exponential survival trees*, Statistics in Medicine **8** (1989), 947 – 961.

[15] Gabriel K. R., *Simultaneous test procedures - some theory of multiple comparisons*, The Annals of Mathematical Statistics **40** (1969), 224 – 250.

[16] Gillo M. W. and Shelly M. W., *Predictive modeling of multivariable and multivariate data*, Journal of the American Statistical Association **69** (1974), 646 – 653.

[17] Kalbfeisch J. D. and Prentice R. L., *The Statistical Analysis of Failure Time Data*, J. Wiley, New York, 1980.

[18] Kass G. V., *An exploratory technique for investigating large quantities of categorical data*, Applied Statistics **29** (1980), 119 – 127.

[19] Loh W. Y. and Vanichsetakul N., *Tree – structured classification via generalized discriminant analysis*, Journal of the American Statistical Association **83** (1988), 715 – 728.

[20] Marubini E., Morabito A. and Valsecchi M. G., *Prognostic factors and risk groups: some results given by using an algorithm suitable for censored survival data*, Statistics in Medicine **2** (1983), 295 – 303.

[21] McCullagh P. and Nelder J. A., *Generalized Linear Models*, Chapman and Hall, New York, 1989. (2nd ed.)

[22] Miettinen O. S., *Theoretical epidemiology: principles of occurrence research in medicine*, J. Wiley, New York, 1985.

[23] Picard C. F., *Théorie des questionnaires*, Gauthiers – Villars, Paris, 1965.

[24] Segal M. R., *Regression trees for censored survival data*, Biometrics **44** (1988), 35 – 48.

[25] Sonquist J. A., Baker E. L. and Morgan J. N., *Searching for structures*, Technical Report, University of Michigan, Institute for Social Research, Ann Arbor, 1973.

[26] Sonquist J. A. and Morgan J. N., *The detection of interaction effects*, Technical Report, University of Michigan, Institute for Social Research, Ann Arbor, 1964.
[27] Stone M., *An asymtotic equivalence of choice of model by cross-validation and Akaike's criterion*, Journal of the Royal Statistical Society **B 39** (1977), 44 – 47.

THE MCGILL UNIVERSITY – MONTREAL CHILDREN'S HOSPITAL RESEARCH INSTITUTE, 2300 RUE TUPPER STREET, MONTRÉAL, QUÉBEC H3H 1P3, CANADA, TEL. (+514) 934 44 00 EXT. 32 88, FAX. (+514) 934 43 31.

E-mail: LOU@BINKLEY.CS.MCGILL.CA or MC39002@MUSICA.MCGILL.CA

STOCHASTIC APPROXIMATION AND OPTIMIZATION

VÁCLAV DUPAČ

Charles University, Prague

ABSTRACT. This is a brief introduction to the topic of stochastic approximation and optimization. An example from the field of water management is given.

Suppose we have to find the solution θ of the equation $f(x) = 0$ where f is a real function of a real variable. One of the commonly used numerical methods of solution is Newton's method described by the iteration

$$x_{t+1} = x_t - (1/f'(x_t))f(x_t).$$

However, this method cannot be made use of in situations when we are able to calculate numerically the function values $f(x)$ only, not the values of the derivative $f'(x)$. For some of these situations a simple iterative formula can be recommended, viz.

$$x_{t+1} = x_t - af(x_t), \tag{1}$$

where a is a constant.

Assume that the graph of the function f lies between two lines of positive slopes passing through the point $(\theta, 0)$. I.e., assume that

$$\operatorname{sign} f(x) = \operatorname{sign}\,(x - \theta),$$

$$K' \mid x - \theta \mid \leq \mid f(x) \mid \leq K'' \mid x - \theta \mid,$$

1990 *Mathematics Subject Classification*. 62L20.

Key words and phrases. Stochastic approximation, mean square convergence of the approximations, stochastic subgradient, penalty paid for the violation of constraints, random inflows.

for all $x \in R$ and some $K' > 0, K'' > 0$. Then the approximations x_t determined by (1) tend to the solution θ for every initial value $x_1 \in R$ provided that $0 < a < 2/K''$.

Consider now the situation when the values $f(x)$ can be found up to an observational error only, i.e. a situation when an observation at point x results in $f(x) + e(x)$ where $e(x)$ is a random noise with zero expectation. We can try to find θ by means of the procedure (1) again, with

$$Y_t = f(X_t) + e_t(X_t)$$

in the place of $f(X_t)$. (Capital letters denote random variables.)

It is easy to see that this time a cannot be chosen as a constant,but has to be replaced by a sequence of positive numbers a_t tending to 0 and such that the sum $\sum_{t=1}^{\infty} a_t$ diverges to infinity. If namely a were a constant for all t, then owing to random errors the differences $X_{t+1} - X_t = aY_t$ could no longer converge to 0 and consequently, nor the sequence of approximations X_t could be convergent. Hence we must have $a_t \to 0$. On the other hand, if the convergence of a_t to 0 were so fast that $\sum_{t=1}^{\infty} a_t = A < +\infty$, then the approximations X_t could tend to another limit than θ. As an example consider $Y(x)$ attaining values $+1, -1$ only, for all $x \in R$, and an initial approximation X_1 at a distance greater than A from θ. An additional assumption $\sum_{t=1}^{\infty} a_t^2 < +\infty$ is usually made which enables utilizing standard probabilistic tools for proving the convergence of the X_t's to θ. We shall formulate a convergence theorem for the multidimensional case right away:

THEOREM. *Let f be a mapping from R^p in R^p, let $\{X_t\}, \{Y_t\}$ be sequences of p-dimensional random vectors. K_0, K_1 etc. will denote positive constants. Let f satisfy the following assumptions: There is a θ such that*

$$| f(x) | \leq K_0 + K_1 | x - \theta |,$$

$$(2) \qquad < f(x), x - \theta > \geq K_2 | x - \theta |^2, \qquad \text{for all} \quad x \in R^p.$$

Writing $Y_t = f(X_t) + e_t, \quad t \geq 1$, we assume

$$(3) \qquad E(e_t \mid X_t) = 0, \qquad E(| e_t |^2 | X_t) \leq K_3 \qquad w.p.\ 1.$$

Choose the constants a_t either as

$$a_t = a/t^{\alpha} \quad \text{with} \quad a > 0,\ 1/2 < \alpha < 1,$$

or as

$$a_t = a/t \quad \text{with} \quad a > 1/2K_2.$$

Then for the sequence

$$X_{t+1} = X_t - a_t Y_t$$

with $E \mid X_1 \mid^2 < +\infty$, *we have*

$$E \mid X_t - \theta \mid^2 = O(a_t).$$

Assumptions of the Theorem can be relaxed; especially, the inequality

$$E(\mid e_t \mid^2 \mid X_t) \leq K_4 + K_5 \mid X_t - \theta \mid^2$$

is sufficient instead of that in (3).

An important application of the described procedure is to finding the point θ at which a real function F of p variables attains its minimal value. The role of f is then played by the gradient of the considered function. However, it is not always possible to estimate the gradient unbiasedly, i.e. with a random error having zero expectation. If only the function values can be estimated unbiasedly (not the values of the gradient), a modified procedure must be made use of. Namely, if $F : R^p \to R^1$ is the function in question and $f : R^p \to R^p$ the gradient, then its ith coordinate is replaced by a difference ratio

$$\text{(4)} \qquad [F(x + cu_i) - F(x - cu_i)]/2c, \quad c > 0,$$

u_i the ith coordinate unit vector $(0, 0, \ldots, 1, \ldots, 0)^{\mathrm{T}}$. Accordingly, Y_t in the iteration scheme

$$X_{t+1} = X_t - a_t Y_t$$

is now defined by

$$(Y_t)_i = [F(X_t + c_t u_i) + e_{2t-1,i} - \{F(X_t - c_t u_i) + e_{2t,i}\}]/2c_t$$

with some constants $c_t > 0$ such that $c_t \to 0$, $\sum a_t c_t < +\infty$, $\sum a_t^2 c_t^{-2} < +\infty$. (If $a_t = a/t$, then e.g. $c_t = ct^{-1/3}$ fulfils these requirements.) The function F is assumed to be twice continuously differentiable. All other assumptions of the Theorem remain unchanged; the assertion is then

$$E \mid X_t - \theta \mid^2 = O(a_t / c_t^2).$$

Under additional assumptions to the Theorem, not only the mean square convergence of the approximations X_t to θ can be proved but also the asymptotic distribution of the normed approximations. With $a_t = a/t, a > 1/2K_2$ the normed approximations $t^{1/2}(X_t - \theta)$ have an asymptotic normal distribution with mean value 0 and covariance matrix given by

$$a^2 \int_0^\infty e^{s(\frac{1}{2}I-aA)} S e^{s(\frac{1}{2}I-aA)^{\mathrm{T}}} \, ds$$

where S is the covariance matrix of the random noise at point θ, A the matrix of first order derivatives of the mapping f at point θ and I the identity matrix.

Consider a mapping $f : R^p \to R^p$ and a point $\theta \in R^p$. Let us recall that the following statements are equivalent:

(i) $$f(\theta) = 0,$$

(ii) $$< f(\theta), x - \theta > \ \geq 0 \quad \text{for all} \quad x \in R^p.$$

Let C be a closed convex subset of R^p. Now,

(i') $$f(\theta) = 0,$$

(ii') $$< f(\theta), x - \theta > \ \geq 0 \quad \text{for all} \quad x \in C$$

are no longer equivalent statements in general. Nevertheless, it is often the θ satisfying (ii') that we are looking for rather than (i'). We shall modify the Theorem for this task.

Assume: C is a closed convex subset of R^p; f is a mapping from C into R^p. $\{X_t\}, \{Y_t\}$ are sequences of p-dimensional random vectors. There is a $\theta \in C$ such that

$$\begin{aligned} < f(\theta), x - \theta > \ &\geq 0 \\ | f(x) | &\leq K_0 + K_1 | x - \theta | \\ < f(x) - f(\theta), x - \theta > \ &\geq K_2 | x - \theta |^2 \quad \textit{for all } x \in C. \end{aligned}$$

Write $Y_t = f(X_t) + e_t$, $t \geq 1$, and assume

$$E(e_t \mid X_t) = 0, \; E(| e_t |^2 | X_t) \leq K_3, \quad t \geq 1.$$

Let constants a_t be as in the Theorem. Denote by π_C the operator of projection onto C. Then, for the sequence

$$X_{t+1} = \pi_C(X_t - a_t Y_t)$$

with $X_1 \in C, E \mid X_1 \mid^2 < +\infty$ we have

$$E \mid X_t - \theta \mid^2 = O(a_t).$$

The Theorem remains true even for set-valued $f : C \to 2^R$ (so-called multifunctions). The assumptions are now to be satisfied by some element of $f(\theta)$ and every element of $f(x)$.

Let us recall a definition: For a real function F of p variables and for every $x \in R^p$ the set

$$\{z \in R^p : F(y) \geq F(x) + < z, y - x >, y \in R^p\}$$

is called the subdifferential of F at x. It is a set-valued function in general; we shall denote it by f again. Obviously, the following statements are equivalent:

$$0 \in f(x) \quad \text{and} \quad \inf\{F(y), y \in R^p\} = F(x).$$

$f(x)$ may be also an empty set; however, if F is convex, then $f(x)$ is not empty.

Let C be a closed convex subset of R^p, let $\theta \in C$. A less trivial equivalence is the following one:

$$F(\theta) = \inf\{F(x), x \in C\} \Leftrightarrow$$
$$\Leftrightarrow < f(\theta), x - \theta > \geq 0 \text{ for all } x \in C \text{ and some element of } f(\theta).$$

Hence, if f is perturbed by random noise, i.e. when only the so-called stochastic subgradient $y(x) = f(x) + e(x)$ can be observed, then the point θ at which function F attains its minimal value on C can be found by iterations according to the modified Theorem (provided all assumptions are fulfilled). The stochastic subgradient may be expressed also in another way as $y(x) = f(x, \omega)$, where ω is a random element such that $Ef(x, \omega) = f(x)$. With a slight inconsistency in notation we use the symbol f for the set (the subdifferential) as well as for its elements (the subgradients).

Still another useful modification of the Theorem

Let C be of the form $C_1 \cap C_2$, where C_1 is a closed convex set and C_2 is defined by

$$\{x \in R^p : F_i(x) \leq 0, 1 \leq i \leq k\}$$

with $F_i : R^p \to R^1$ convex, differentiable and such that

$$\{x \in C_1 : F_i(x) < 0, 1 \leq i \leq k\} \neq \emptyset.$$

A function F is to be minimized on C. Its subdifferential will be again denoted by f. It is assumed that

$$< f(\theta), x - \theta > \; \geq 0$$
$$| f(x) | \leq K_0 + K_1 | x - \theta |$$
$$< f(x) - f(\theta), x - \theta > \; \geq K_2 | x - \theta |^2$$

for all $x \in C$, some element of $f(\theta)$ and every element of $f(x)$. Denote $H(x) = \sum_{i=1}^{k} F_i(x)^+$, $h(x)$ the subdifferential of H. Let a_t be as in the Theorem. Let $\{X_t\}, \{Y_t\}, \{U_t\}$ be sequences of p-dimensional random vectors. Write $Y_t = f(X_t) + e_t, U_t = h(X_t) + e_t^*$. Assume

$$E(e_t \mid X_t) = 0, \;\; E(e_t^* \mid X_t) = 0,$$

$$E(| e_t |^2 | X_t) \leq K_3, \;\; s_t^2 E(| e_t^* |^2 | X_t) \leq K_4,$$

where $\{s_t\}$ is a sequence of nonnegative constants such that $s_t \to \infty$, $a_t s_t^2 \to 0$. Then for the sequence

$$X_{t+1} = \pi_C(X_t - a_t(Y_t + s_t U_t)), \;\; t \geq 1,$$

with $E | X_1 |^2 < \infty$, we have

$$E | X_t - \theta |^2 = O(a_t).$$

Here, $H(x)$ can be interpreted as a global size of violation of the constraints $F_i(x) \leq 0$, $1 \leq i \leq k$, and $s_t H(X_t)$ as a penalty paid for this violation.

If the subgradients f and h cannot be estimated unbiasedly, then we may again, as in (4), replace $f(X_t)_i$ by

$$[F(X_t + c_t u_i) - F(X_t - c_t u_i)]/2c_t;$$

similarly $h(X_t)_i$. In this case, however, some additional assumptions are needed to prove convergence.

Several modifications of the described procedures have been proposed. Especially, the constants a_t may be chosen not apriori but adaptively, according to the evolution of the sequence of approximations. The a_t's can be kept large as long as the $X_t's$ approach to the solution monotonously, and made smaller, when the X_t's begin to oscillate; this intervention can be done repeatedly.

An example of the use of the procedure:

A water reservoir on a river is to be built. Its main purposes are:

- irrigation water supply
- water withdrawal for a thermal power plant and other industrial uses
- flood control
- recreation and fishery.

The task is to minimize the cost of the reservoir under condition that the listed needs will be met at least with a prescribed probability. As the cost is an increasing function of reservoir capacity it suffices to minimize the capacity.

The hydrological year, starting with November, is (for these purposes only) divided into four periods:

i) November till April of the following year
ii) May and June
iii) July and August
iv) September and October.

Industrial demands enter into the deterministic (nonrandom) constraints, the water supply for them must not fall down under the levels $d_i, 1 \leq i \leq 4$, in the four periods of the year. The irrigation water supply is required only in the second, third and fourth period and is viewed as a random vector $(\beta_2, \beta_3, \beta_4)$ with a known 3-dimensional normal distribution. The enviromental, fishing and recreation needs imply some minimal reservoir storage for all four periods and the flood control target requires some prescribed storage be free in the reservoir operation during all periods. The cumulated water inflows in the four periods are supposed to lie within some known probabilistic bounds (appropriate quantiles).

Let us denote x_i, $1 \leq i \leq 4$, the released water volumes in the four periods and denote x_0 the reservoir capacity. There are upper bounds for the x_i's and for x_0 which stem from natural hydrological situation. The problem can be then formulated as follows:

Minimize x_0

subject to

$$P(x_i \geq \beta_i + d_i,\ i = 2,3,4) \geq \alpha \tag{5}$$

and subject to deterministic constraints of the type

$$\sum_{i=1}^{k} x_i \leq b_k, \quad 1 \leq k \leq 4$$

$$\sum_{i=0}^{k} x_i \geq c_k, \quad 1 \leq k \leq 4$$

$$l_i \leq x_i \leq u_i, \quad 0 \leq i \leq 4$$

The constraints define the closed convex set C. The constants b_k, c_k, l_i, u_i can be easily calculated, when all the above mentioned restrictions as well as the probability distributions of random inflows have been specified.

For a direct application of the approximation procedure, the task is to be reformulated as a minimization of a function $F(x)$, $x = (x_0, x_1, x_2, x_3, x_4)$, whose values can be unbiasedly estimated, that is, whose values are expected values of observations. In our example, the "observations" will be based on simulated values of the random element $\omega = (\beta_2, \beta_3, \beta_4)$. A possible reformulation of the problem is the following:

Find the point θ *at which the function*

$$F(x) = x_0 + c\{\alpha - E(I[x_i \geq \beta_i + d_i,\ i = 2,3,4])\}$$

attains its minimal value on the set C. *Here* I *is the indicator of the event in brackets and c is a very large constant, that could be also replaced by a sequence* $c_t \to \infty$ *in the course of the procedure.*

In fact, however, a bit different approach has been adopted:

Minimize on the set C the function

$$F(x) = x_0 + cE\{\max(\beta_i + d_i - x_i,\ i = 2,3,4)^+\}$$

with $c > 0$ moderately large. After the solution $\theta = (\theta_i,\ i = 0,1,2,3,4)$ has been found,the value of the probability (5) is calculated for $x = \theta$. If it is smaller than α, a new, larger $c > 0$ is chosen and the process repeated. If, on the other hand, the probability is much higher than required, a smaller $c > 0$ can be chosen.

Stochastic subgradient $y(x)$ of the function $F(x)$ can be calculated here; we get

$$f(x,\omega)_0 = 1,\ f(x,\omega)_1 = 0$$

and for $i = 2,3,4$

$$\begin{aligned} f(x,\omega)_i &= -c, \quad \text{if } \beta_i + d_i - x_i = \max(\beta_i + d_i - x_i,\ i = 2,3,4)^+, \\ &= 0 \quad \text{otherwise.} \end{aligned}$$

Alternatively, if computationally preferable, the ith coordinate of the stochastic subgradient $f(x,\omega)_i,\ i = 2,3,4$, can be estimated by

$$c\{\max(\beta_i + d_i - (x_i + \varepsilon u_i),\ i = 2,3,4)^+ - \\ - \max(\beta_i + d_i - (x_i - \varepsilon u_i),\ i = 2,3,4)^+\}/2\varepsilon$$

with u_i the ith coordinate unit vector and $\varepsilon > 0$ as small as computationally possible.

A detailed and lucid exposition of stochastic approximation and optimization is given in Schmetterer [1]. Our example is taken from Dupačová et al [2]. Further examples as well as theoretical and computational developments can be found in the collection of papers edited by Ermoliev and Wets [3].

REFERENCES

[1] Schmetterer, L., *From stochastic approximation to the stochastic theory of optimization, 11. Steiermärk. Math. Symp., Stift Rein*, Bericht Nr. 127 der Mathematisch-Statistischen Sektion im Forschungszentrum Graz, 1979.

[2] Dupačová J., Gaivoronski A., Kos Z. and Szantai T., *Stochastic programming in water management: A case study and a comparison of solution techniques*, European Journal of Operational Research **52** (1991), 28-44.

[3] Ermoliev Yu. and Wets R. (eds.), *Numerical Techniques for Stochastic Optimization Problems*, Springer, Berlin, 1988.

CHARLES UNIVERSITY, DEPARTMENT OF PROBABILITY AND MATHEMATICAL STATISTICS, SOKOLOVSKÁ 83, CS – 186 00 PRAHA 8, CZECHOSLOVAKIA; TEL: (+422) 231 78 62; FAX: (+422) 84 76 88; TELEX: 121 673 MFF.

E-mail: DUPACV@CSPGUK11.BITNET

COMPARISON OF THE STOCHASTIC APPROXIMATION SOFTWARE

PAVEL CHARAMZA
Charles University, Prague

ABSTRACT. In this short review the available PC software for stochastic optimization based on the recursive estimation is described. Especially, Stochastic quasigradient software SQG-PC and Stochastic nonlinear problem software SNLP are compared. In the last section problems of possible 'outliers' in the optimization problems are considered and their influence to the behavior of recursive process. Some hints are given how to overcome this outlier difficulties together with software recommendation.

1. STOCHASTIC QUASIGRADIENT SOFTWARE - SQG PC.

SQG PC was developed in the International Institute for Applied System Analysis (IIASA) in Laxenburg, Austria. A detailed description is given in Gaivoronski [1988]. Here we mention only the main features of this software. The software was developed in IBM Fortran 77, v. 1.22, hence a coprocessor and an IBM Fortran (RM Fortran) compiler are needed for its performance. It works on PC's compatible with IBM.

Problem 1.1. *The problems which can be solved by SQG are those of minimizing the function $Ef(x, \omega)$ over $x \in X$, where E denotes the expectation and ω is a random variable. x is a vector from a subset X (feasible set) of the m-dimensional Euclidean space and f is a real valued function (objective function) of x and ω. X is supposed to be specified by a system of general linear constraints and bounds.*

The algorithm used for solving Problem 1.1 is based on the following recursive scheme.

1990 *Mathematics Subject Classification.* 90C15, 94-15.
Key words and phrases. stochastic optimization, stochastic approximation, stochastic quasigradients, outliers, robustness.

Algorithm 1.1.

$$x(s+1) = \prod_X (x(s) - \rho(s)\xi(s)),$$

where $x(s)$ is the estimate of the solution at the s-th step, $\rho(s)$ is the stepsize of the algorithm at the same step and $\xi(s)$ is a random vector for which $E\xi(s) = E\nabla f(x,\omega) + a(s)$. Here ∇ is a subgradient (gradient) operator and a(s) is a small vanishing term in the limit. $\xi(s)$ is called a stochastic quasigradient (see, e.g., [Ermoliev 1988] for exact definitions). (By $\prod_X$ we denote the operator of projection onto the set X.)

The more than sufficient theoretical results are known about the asymptotic properties of $x(s)$. An overlook of these properties can also be found in this book, see Dupač [1992].

1.1. Input parameters.

The input of the SQG software can be divided in two categories:

- input parameters which must be prepared before running the software (objective function, random numbers generator, definition of the set X, initial point $x(0)$.
- input parameters which can be adaptively changed in the course of the SQG program.

Let us consider the first category inputs. We have to define the function f. Since SQG does not have an algebraic analyzer we have to define f in advance as a FORTRAN function, compile it and link to the main already compiled body of the SQG. If we know $E\nabla f(x,\omega)$ in advance (i.e. if $a(s) = 0$) then we can define it as a FORTRAN procedure. We have to compile it and link together with the objective function and the main object SQG module. Otherwise, $E\nabla f(x,\omega)$ is estimated from the values of f using finite differences scheme. There are several possibilities how to choose these finite differences schemes:

(1) *central differences scheme* (e(i) denotes i-th unit vector)

$$(2\delta)^{-1} \sum_i (f(x(s) + \delta e(i), \omega_{1i}) - f(x(s) - \delta e(i), \omega_{2i}))e(i)$$

(2) *forward differences scheme*

$$\delta^{-1} \sum_i (f(x(s) + \delta e(i), \omega_{1i}) - f(x(s), \omega_{2i}))e(i)$$

(3) *random search* ($t(i)$ is a vector with idependent components uniformly distributed on the interval $(0, \delta)$)

$$\sum_i \| t(i) \|^{-1} (f(x(s) + t(i), \omega_1) - f(x(s), \omega_2))\, t(i).$$

If the user does not define the procedure for a (sub)gradient computation, the type of a difference scheme is then determined after running the program itself.

The random number generator is also defined by the user as a FORTRAN function which can be used in the source code for the objective function and (sub)gradient procedure. This generator function must be compiled and linked together with objective function and (sub)gradient procedure source codes and main SQG object code.

Before running the program which is now created the set X of constraints must be defined. This is especially true if X is a general linear constraints set. The matrix and the right hand side values of the constraints must be defined in an ASCII file in SQG-PC format which is quite simple. We are not going into its details here. If the constraints are of a simple bound type (the coordinates of x are in some deterministic intervals) then the feasible set can be defined after running the program. This is also possible if the feasible set consists of a single linear constraint.

The initial iteration point $x(0)$ must be also predetermined in an ASCII file before starting the program.

The input parameters of the second type can be changed after running the program itself. This means that the user can produce several processes of the type defined in Algorithm 1.1 with different parameters. Moreover, these parameters can be changed during the process itself, adaptively, according to the behaviour of the optimization process. The parameters that can be affected are: the type of stepsize $\rho(s)$ computation (constant, three types of adaptive stepsize, standard type of $c1/(s + c2)$, some kind of variable metrics type), direction $\xi(s)$ computation (FORTRAN procedure defined by the user or some of the finite difference schemes, averaging principle possibility for subgradient estimations, some normalization features). If the set X is of a simple type (simple deterministic bounds for any of the coordinates of x, one linear constraint) then the constraints can be also changed adaptively during the program run. It should be mentioned that the projection $\prod$ from the algorithm 1.1 is not computed exactly (since this is a special quadratic programming problem consuming a lot of PC's time) but it is only approximated by another recursive deterministic scheme. The user can also change the type of function value $Ef(x(s), \omega)$ estimate and he has some other possibilities not so frequently used to be mentioned in this short overview.

1.2 Output of SQG.

There are three kinds of output - graphical, numerical and disk file output. Since the program was developed several years ago the graphical output is designed for CGA graphics cards only. Nevertheless, it can be used on any graphics card of the higher level (EGA, VGA), and its quality is sufficient for basic information concerning the optimization process behaviour and for practical computation purposes. Zooming, change of scale, automatic scaling etc. are the positive features of the graphical output. For all the three types of output you can choose the coordinates of x which you would like to demonstrate. You can also demonstrate the estimates of the function value $Ef(x, \omega)$ based on the different type of estimates (averages based on the last k observations, averages based on simulations, etc. If you prefer not adaptively controlled (but quick) computation you can also switch on the output only for the last iteration demonstration. This can be used, e.g., when you have already tuned the parameters of the process and you would like to compute the final result.

2. Stochastic nonlinear program - SNLP.

SNLP software is introduced in this book, c. f. Loskutov [1992].

3. Comparison of SQG and SNLP.

In this chapter we give basic comparison of SQG and SNLP. In Table 3.1 we show which standard optimization problems we are able to solve using SQG and SNLP. In the column **choice** we give our recommendation which software is better to use for that particular optimization problem.

optimization problem	SQG	SNLP	choice
linear programming simplex method	no	yes	SNLP
nonlinear programming	yes	yes	BOTH
stochastic programming linear constraints	yes	yes	SQG
simple linear recourse	yes	yes	SNLP

Table 3.1

The recommendation from the column **choice** should be taken into account together with the following Table 3.2. In this table we summarize our experience with using this software. Since we use this software also for teaching purposes, we give also our opinion of a possible use of SQG and SNLP in educational process.

Question	**SQG**	**SNLP**
Must one be an expert in optimization for using this software?	good expert	medium expert
Must one be an expert in computers for using this software?	medium expert IBM Fortran	medium expert MS Fortran
Is it possible to control the computational proccess?	good control	worse control
Is it easy to handle it?	menu driven system	menu driven system
Is there a good description?	excellent	good
Can we use it for basic optimization courses?	no	sometimes
Can we use it for advanced optimization courses?	yes	yes
Is it possible to use it for practical purposes?	sure	sure

Table 3.2

The main disadvantage of both software packages is that they do not have algebraic analyzer for function, its gradient and random numbers input. This means that if you want to use SQG or use SNLP for nonlinear optimization then you have to create your own FORTRAN source code, compile it and link together with the main object code. In SNLP there are some basic random generators (normal, uniform, exponential) which can be used from 'inside' of the program without 'outside' compiling, nevertheless, nonlinear problems or general distributions must be input from outside. Linear program solver (simplex algorithm) is available from inside of SNLP, i.e. no preliminary compilations are needed.

4. Outliers in stochastic optimization.

If you want to solve Problem 1.1 it is usual in applications that you know the distribution of ω in advance. This distribution can be

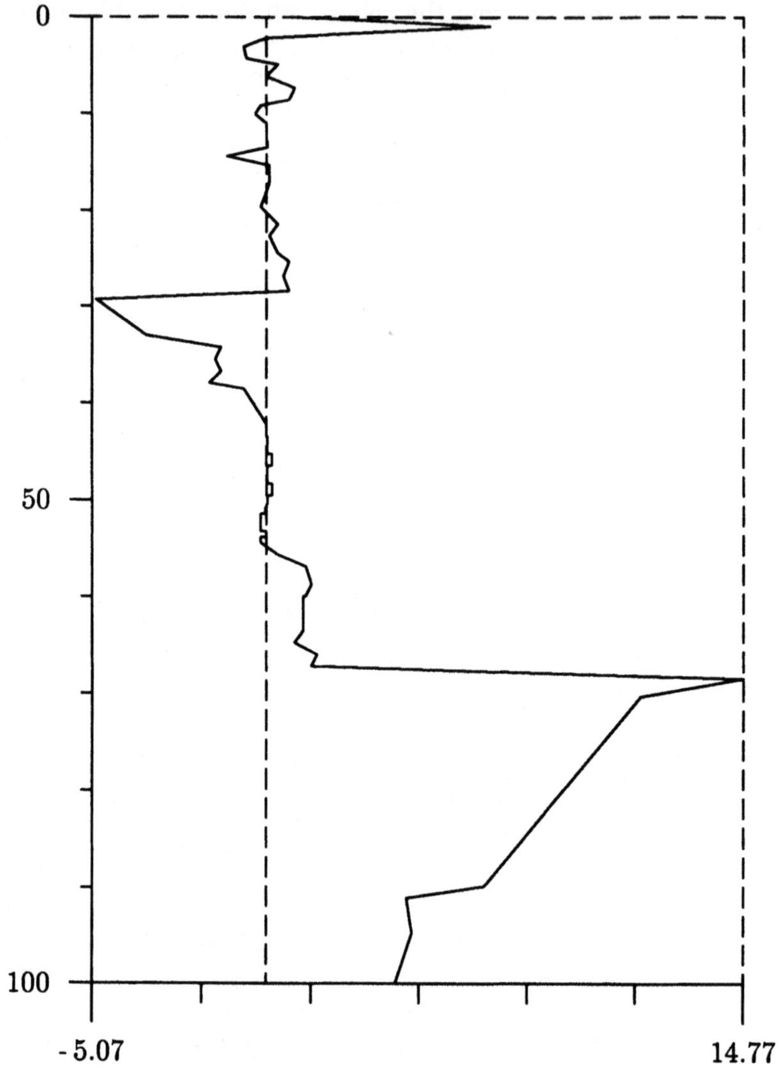

Picture 4.1. Severe irregularities of the process $x(s)$, where the minimum of $f(x, \omega) = -x^2 + \omega$ was to be found.

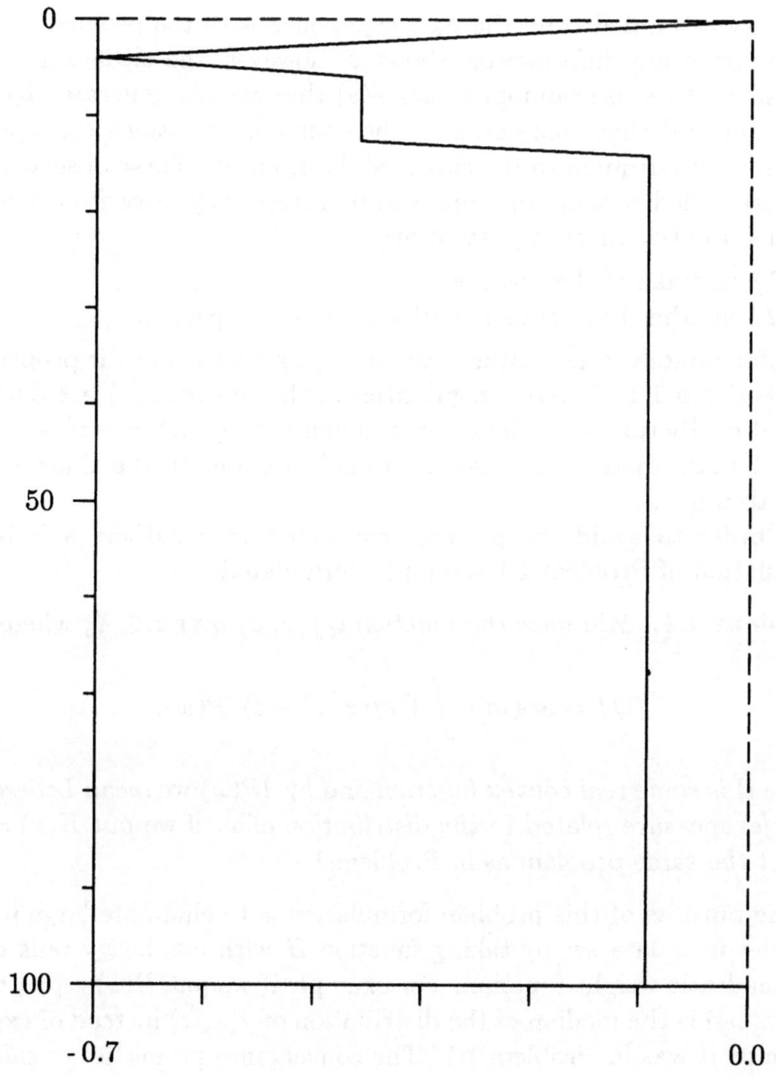

Picture 4.2. The behaviour of the Algorithm 4.2 for finding the root of the equation $Q(-2 * x + \omega) = 0$, with Q the median of the distribution of ω which is again the Cauchy one.

somehow estimated from the past experience with the problem. If we do not have any information about ω, however, the situation is not the same. If we use random values $\omega(s)$ that are not generated by the computer and that come on-line, then some observations can appear that are not adequate to the expected distribution. These observations that are called outliers in statistical literature may arise from several reasons, the two most important are:

(1) mistake of the observer;
(2) another type of distribution then it is expected.

Unfortunately, these outliers can damage good asymptotic properties of Algorithm 1.1. Severe irregularities of the process $x(s)$ are demonstrated on Picture 4.1, where the minimum of $f(x,\omega) = -x^2 + \omega$ was to be found. Instead of a 'good normal' distribution the Cauchy one was taken for ω.

In order to avoid the problem connected with outliers a 'robust' formulation of Problem 1.1 should be formulated.

Problem 4.1. *Minimize the function $Qf(x,\omega)$ over $x \in X$, where*

$$Qf = \arg\min_t \int H(f(x,\omega) - t)dP(w),$$

where H is some real convex function and by $dP(\omega)$ we mean Lebesque-Stieltjes measure related to the distribution of ω. If we put $H(z) = z^2$ we get the same problem as in Problem 1.1.

The purpose of this problem formulation is to eliminate large irregularities in a data set by taking function H with less heavy tails then the quadratic weight function. For example if we put $H(z) = \mid z \mid$ then $Q(f(x,\omega))$ is the median of the distribution of $f(x,\omega)$ instead of expectation as it was in Problem 1.1. The convergence properties of sample median compared with sample average in the presence of outliers are well known from the basic courses of statistics. Hence we intuitively expect better convergence properties if we solve Problem 4.1 instead of Problem 1.1. Although a solution of Problem 4.1 can differ from that of Problem 1.1 the intuitive relations between these two problems are evident. However, algorithms for solving 4.1 have not been developed that extensivelly yet. Here we mention briefly the one dimensional case only.

Algorithm 4.1 *(hint). Let us suppose that we are at the s-th iteration, i.e. we have already got points $x(0), x(1), \ldots, x(s)$ (that are real numbers now). Let us assume that the $x(i)$'s are mutually different. At* *each of these points we have 'observations'* $f(x(i),\omega(i)), i = 0,\ldots,s$. *Now,*$x(i), i = 1,\ldots,s$ *are real numbers. Let* $y(0),\ldots,y(s)$ *are the values of* $x(i)$*'s ordered in a nondecreasing sequence. We find the nondecreasing function g which is defined at points* $x(i), i = 1,\ldots,s$ *such that it minimizes* $\sum_{t=1}^{s} H(f'(y(t)) - g(y(t)))$, *where* $f'(y(t)) = f(y(t),\omega(i_t)) - f(y(t-1),\omega(i_{t-1}))$. *(*$i_t$ *is the index of* $x(i_t)$ *which corresponds to* $y(t)$*). We put* $x(s+1)$ *as the centre of* $< y(imin), y(imax) >$, *where imin and imax are adjacent indeces and* $g(y(imin)) <= 0, g(y(imax)) >= 0$.

A more accurate description with convergence properties is given in Charamza [1989].

Another possible approach to eliminating the influence of outliers is based on differentiating f and finding the 'root' of its derivative. The problem of minimizing can be transformed into the problem of finding a solution of systems of nonlinear equations in a deterministic case. Taking Problem 1.1 and interchanging the expectation operator with the differentiating operation we get the problem of finding the solution of nonlinear equations $E\nabla f(x,\omega) = 0$ instead of Problem 1.1. Taking another measure of location parameter than the expected value we can formulate the alternative problem to 4.1.

Problem 4.2. *Find a solution of the system of equation* $Q\nabla f(x,\omega) = 0$, *over* $x \in X$. *Since Q was defined for random variables only we take for Q of a vector* ∇f *the vector of Q of its coordinates. That means if H from Problem 4.1 is the quadratic function then Q is the multidimensional expectation.*

There is no theory known about the theoretical relations between the solution of Problem 1.1 and Problems 4.1, 4.2 respectively. However, convergence results of the solutions of empirical versions of Problem 4.2 to the solutions of the corresponding theoretical problem are known at least in one dimensional case. Also the practical experience is good in this case. For this reason we introduce the algorithm for estimating the solution of Problem 4.2 in one dimensional case. It is very similar to the Algorithm 4.1.

Algorithm 4.2 *(hint). Let us suppose that we are at the s-th iteration, i.e. we have got points* $x(0), x(1),\ldots,x(s)$. *Without loss of generality let us assume that* $x(i)$*'s are mutually different. At each of these points we have 'observations'* $f(x(i),\omega(i)), i = 0,\ldots,s$. *Let* $y(0),\ldots,y(s)$ *are the values of the* $x(i)$*'s ordered in a nondecreasing sequence. We find the nondecreasing function g which is defined at the points* $x(i), i = 1,\ldots,s$ *such that it minimizes* $\sum_{t=0}^{s} H(f(y(t),w(i_t)) - g(y(t)))$. *We put* $x(s+1)$

as the centre of

$$< y(imin), y(imax) >,$$

where imin and imax are adjacent indeces and

$$g(y(imin)) \leq 0, \quad g(y(imax)) \geq 0.$$

A more detailed description of this algorithm together with an algorithm for finding isotonic function g is given in Charamza [1989a], Charamza [1989b].

On picture 4.2 we show the behaviour of the Algorithm 4.2 for finding the root of the equation $Q(-2 * x + \omega) = 0$, with Q the median of the distribution of ω which is again the Cauchy one. Comparing this behaviour with that from Chapter 1 we can see a better result.

There exists a software *Stochastic Approximation* which gives the possibility of comparison of different types of algorithms for solving problems of the type 1.1, 4.1, 4.2. The standard statistical estimation problems of location parameters, quantiles, LD-50 quantities and other estimates using recursive schemes of the type of Algorithms 1.1, 4.1, 4.2 can be solved using this software, too. Here we give a brief list of methods available in this software. The full description of the program which can be obtained free from the author is given in Charamza [1992]. The main disadvantage of this software is its one dimension limitation, i.e. it can be used only for comparisons of large variety of different types of stochastic approximation algorithms but not for a practical optimization. However, it can be useful for solving some basic sequential statistical estimates problems.

List of algorithms available in Stochastic Approximation Package.

A. Problem of finding the root of a regression function.

The problem can be formulated as solving the equation

$$Eg(x, \omega) = \alpha, \tag{4.1}$$

where g is a real valued function and α is some given real number which can be chosen arbitrarily in the program. The function g can also be defined by the user.

Methods available for solving problem (4.1):

(1) Robbins–Monro procedure, c. f. Robbins-Monro [1951], Blum [1954];

(2) Adaptive Robbins-Monro procedure, c. f. Fabian [1968];
(3) Procedure based on isotonic regression, c. f. Mukerjee [1981], Charamza [1984];
(4) Procedure based on quasiisotonic regression, c. f. Dupač [1987];
(5) Procedure based on L1-quasiisotonic regression, c. f. Charamza [1990];
(6) Pflug procedure, c. f. Pflug [1988].

B. Problem of finding the root of a regression quantile function.

The procedures of this paragraph solve iteratively the equation

$$Q(x,\gamma) = \alpha, \tag{4.2}$$

where $Q(x,\gamma)$ is defined as the γ–*quantile* of the distribution according to which the observations at the point x are obtained.

(1) Procedure based on isotonic quantile regression, c. f. Charamza [1990];
(2) Procedure based on quasiisotonic quantile regression, c. f. Charamza [1990];
(3) Derman procedure, c. f. Derman [1954];
(4) Procedure based on isotonic regression, c. f. Mukerjee [1981], Charamza [1990];
(5) Blum procedure, c. f. Blum [1954];
(6) Bather procedure, c. f. Bather [1990].

C. Problem of finding the root of a regression location parameter function.

Let us denote by F_x the distribution function of observations provided at the point x. Let $H : (R^2, B^2) \to (R, B)$. By the regression location parameter function we mean the real measurable function $m(x)$ defined by

$$m(x) = \arg\min_{\theta} \int H(y,\theta) dF_x(y). \tag{4.3}$$

The methods listed below solve iteratively the equation

$$m(x) = \alpha. \tag{4.4}$$

(1) Procedure based on isotonic regression, c. f. Charamza [1990];
(2) Procedure based on quasiisotonic regression, c. f. Charamza [1990].

D. Estimation of the extreme of a regression function.

The methods from this paragraph are designed for finding the extreme of the function $m(x) = Eg(x,\omega)$, where the function g is the same as in (4.1).

(1) Kiefer-Wolfowitz method, c. f. Kiefer-Wolfowitz [1952];
(2) Isotonic regression for differences, c. f. Charamza [1984];
(3) Quasiisotonic regression for differences, c. f. Charamza [1990].

Estimation of the quantile.

The procedures in this paragraph can find the γ-quantile of an unknown distribution from a random sample. All the methods are recursive.

(1) Tierney method, c. f. Tierney [1981];
(2) Procedure based on quasiisotonic regression, c. f. Charamza [1990];
(3) Procedure based on isotonic regression, c. f. Charamza [1990];
(4) Derman procedure; Derman [1957].

F. LD-50 problem.

Let us have an unknown distribution function F. The problem is to find the $\gamma - quantile$ of this distribution. In comparison with the problem formulated in the preceeding paragraph we do not have a random sample from this distribution at our disposal. The only information that we can obtain about this distribution is given by random variables $Y(X(t))$ distributed according the 0-1 law:

$$\begin{aligned} Y(X(t)) = 0 \quad & \textit{with probability} \;\; 1 - F(X(t)) \\ Y(X(t)) = 1 \quad & \textit{with probability} \;\; F(X(t). \end{aligned}$$

(1) Procedure based on the isotonic regression, c. f. Charamza [1990];
(2) Derman procedure, c. f. Derman [1957].

References

[1] Barlow R .E., Bartholomew D. J., Bremner J. M. and Brunk H. D., *Statistical Inferennce Under Order Restrictions*, Wiley, London, 1972.

[2] Bather J. A., *Stochastic approximation: A generalization of the Robbins–Monro procedure, In: Proceedings of the Fourth Prague Symposium on Asymptotic Statistics* (Mandl P. and Hušková M., eds.), Charles Univ. Publ. House, 1989.

[3] Blum J. R., *Multidimensional stochastic approximation procedures*, Ann. Math. Statist. (1954), 737 – 744.

[4] Charamza P., *Stochastic approximation on a lattice*, *Diploma Work*, Fac. of Mathematics and Physics, Prague, 1984. (in Czech)

[5] Charamza P., *Non – standard stochastic approximation scheme*, IIASA Working Paper (1989a).

[6] Charamza P., *Robust isotonic regression.*, *In: Proceedings of the 6th European Young Statisticians Meeting* (Hála M. and Malý M., eds.), Charles Univ. Prague, 1989b.

[7] Charamza P., *Isotonic regression and stochastic approximation.*, Thesis on the Faculty of Math. and Physics (1989c), Charles Univ. Prague, Prague. (in Czech)

[8] Charamza P., *Integrated stochastic approximation program system*, In Lecture Notes in Economics and Mathematical Systems **374** (1992) (Pflug G., ed.), Springer.

[9] Derman C., *Non – parametric up – and – down experimentation*, Annals of Math. Statistics **28** (1954), 795 – 797.

[10] Dixon and Mood, *A method for obtaining and analyzing sensitivity data*, J. Ammer. Stat. Assoc. **43** (1948), 109 – 126.

[11] Dupač V., *On stochastic approximation*, Kybernetika **17 (Appendix)** (1981). (in Czech)

[12] Dupač V., *Stochastic approximation.*, *In: Handbook of Statistics*, (Krishnaiah P.R., Sen P.K., eds.), vol. 4, Elsevier Science Publ., Amsterdam, 1984, pp. 515 – 529.

[13] Dupač V., *Quasi – isotonic regression and stochastic approximation*, Metrika **34** (1987), 117 – 123.

[14] Dupač V., *Stochastic approximation and optimization*, *This proceedings.*

[15] Ermoliev Yu. M., *Methods of Stochastic Programming*, Nauka, Moscow, 1976. (in Russian)

[16] Ermoliev Yu. M., *Stochastic quasigradient methods and their application in systems optimization*, IIASA Working Paper – 81 – 2 (1981).

[17] Gaivoronski A., *Interactive program SQG – PC for solving stochastic programming problems on IBM PC/XT/AT compatibles – user guide*, *In:Numerical Techniques for Stochastic Optimization* (Ermoliev Yu. and Wets R., eds.), Springer – Verlag, 1988.

[18] Huber P. J., *The behavior of MLE under non standard conditions*, *In: Proceedings of the Fifth Berkeley Symposium on Math. Statistics and Prob. Theory*, Univ. of California Press..

[19] Hušková M., *Recursive M – test for detection of change*, Sequential Analysis **7** (1988), 75 – 90.

[20] Kiefer J. Wolfowitz J., *Stochastic estimation of the maximum of a regression function*, Ann. Math. Statist. **23** (1952), 462 – 466.

[21] Kirchen A., *Ûberlegungen zur eindimensionalen stochastischen approximation*, *Diploma work on University Bonn*, 1982.

[22] Loeve M., *Probability Theory*, Van Nostrand, New York, 1963.

[23] Loskutov V., *Models, algorithms and software of stochastic optimization*, *This proceedings*, 1992.

[24] Loskutov V., Roenko N., Uryasiev S., *Stochastic Nonlinear Programming System*, IIASA Working Paper – 89 – 075. (1989).

[25] Menendez and Salvador, *An algorithm for isotonic median regression*, Computational Statistics and Data Analysis **5** (1987), 399 – 406.

[26] Mukerjee, H. G., *A stochastic approximation by observation on a discrete lattice using isotonic regression*, Ann. Math. Statistics **9** (1981), 1020 – 1025.

[27] Nemirovski, A. S., Polyak, B. T., Cybakov A. B., *Signal processing using non-parametric maximum likelihood method*, Problemy peredaci informacii **20** (1984). (in Russian)

[28] Nevelson and Khasminskij, *Stochastic Approximation and Recursive estimation*, Nauka, Moscow, 1972. (in Russian)

[29] Ermoliev Yu. M. and Wets R. (eds.), *Numerical Techniques for Stochastic Optimization*, Springer – Verlag, 1988.

[30] Pflug G., *Stepsize rules, stopping times and their implementation in stochastic quasigradient methods, In: Numerical Techniques for Stochastic Optimization* (Ermoliev Yu.M., Wets R., eds.), Springer- Verlag, 1988.

[31] Robbins H. and Lai T. L., *Consistency and asymptotic efficiency of slope estimates in stochastic approximation schemes*, Z. Wahrsch. Verw. Gebiete **56** (1981), 329 – 360.

[32] Robbins H. and Monro S., *A stochastic approximation method*, Annals Math. Statistics **22** (1951), 400 – 407.

[33] Roenko N. (see Loskutov V.).

[34] Schmetterer L., *From stochastic approximation to the stochastic theory of optimization*, Ber. Math. – Statist:Sekt. Forsch. Graz, number 27 (1979).

[35] Tierney L., *A space – efficient recursive procedure for estimating a quantile of an unknown distribution*, SIAM J. Sci. Stat. Comput. **4** (1983).

[36] Uryasiev S. (see Loskutov).

[37] Wasan M.T., *Stochastic Approximation*, Cambridge University Press, 1969.

[38] Wichman P. and Hill D., *Building a random number generator*, Byte **12**, 127 – 128.

CHARLES UNIVERSITY, DEPARTMENT OF PROBABILITY AND MATHEMATICAL STATISTICS, SOKOLOVSKÁ 83, CS – 186 00 PRAHA 8, CZECHOSLOVAKIA; TEL: (+422) 231 78 62; FAX: (+422) 84 76 88; TELEX: 121 673 MFF.

E-mail: CHARAMZA@CSPGUK11.BITNET

MODELS, ALGORITHMS AND SOFTWARE OF STOCHASTIC OPTIMIZATION

VICTOR LOSKUTOV

Institute of Cybernetics, Kiev

ABSTRACT. The main objective of this paper is to consider stochastic generalization of linear and nonlinear optimization problems, methods of solution and relevant software.

1. INTRODUCTION

The family of software for solving linear and nonlinear programming, stochastic optimization problems and relevant scientific background are considered.

This family was developed above all out of necessity to solve practical problems. These are problems of power systems development accounting for device failures, problems of prospective planning accounting for demand uncertainty and agricultural problems accounting for weather conditions. Taking into consideration that the main source of stochastic problems is represented by deterministic problems, our software provides for solving both stochastic and deterministic problems. It also supports the conversion of a deterministic problem into its stochastic analog and vice versa.

There is no exaggeration in stating that two-stage stochastic problems, or problems with recourse, occupy a central place among the range of stochastic optimization problems. Evidence of this is found in the voluminous literature dedicated to these problems, and in the activity in the development of software for their solution. The most widely spread are linear stochastic problems with recourse, which may be regarded as a certain generalization of linear (LP) and nonlinear

1990 *Mathematics Subject Classification*. 90C15, 94-15.

Key words and phrases. Linear programming, nonlinear programming, stochastic optimization, stochastic quasigradients.

(NLP) programming problems with linear constraints accounting for the possible uncertainty of the source data.

As a rule, the values of many parameters in real problems are never absolutely accurate. This inaccuracy may have various nature. For instance, in problems of long-term planning some model coefficients are obtained by forecasting, and in the macroeconomic models some coefficients are obtained by means of aggregate evaluation. Among the sources of the above inaccuracy may be impossibility or extreme sophistication of the precise evaluation of parameters, use of expert estimates concerning the parameter values, etc. For certain types of problems, e g. the agricultural production ones, this inaccuracy of parameters may amount to 100% and more.

Disregarding the uncertain nature of parameters and replacing them by specific values (for instance, mathematical expectation) in the solution of LP or NLP problems may lead to invalid results, due to the problem's instability in relation to the source data. Consequently, the optimization model may be inadequate. Information about the solution stability of a problem may be obtained by means of post-optimum LP-analysis; the latter's abilities, however, are rather limited.

If this adequacy is lacking, the given situation may be tackled by various approaches: optimization by minimax criterium, converting a problem to a stochastic programming one with probability constraints, or to a stochastic problem with recourse, see [1].

The software discussed deals with the latter approach based upon creating the stochastic analog for a linear and nonlinear programming problem, i. e stochastic problem with recourse. This approach assumes that uncertain or inaccurate source data of the model may be interpreted as random values.

The main ideas used in the algorithm for solving stochastic problems are reflected in [2]. A special modification of the so called stochastic quasigradient algorithms for stochastic problems with recourse is implemented.

2. PROBLEMS STATEMENT

The following problems are considered:

- linear programming problems;
- nonlinear programming problems under linear constraints; in particular problems of nondifferentiable optimization;
- stochastic programming problems, in particular problems with recourse.

The nonlinear programming problems are following, i. e.

$$\min\ F(x) \tag{1}$$

subject to

$$l \leq \begin{pmatrix} x \\ Ax \end{pmatrix} \leq u. \tag{2}$$

If the function $F(x)$ is linear, that is $F(x) = cx$, the problem turns into a linear programming problem. If the function $F(x)$ is nonsmoth, it is a problem of nondifferentiable optimization. If the function $F(x)$ is mathematical expectation of stochastic function $f(x, w)$

$$F(x) = E\, f(x, w), \tag{1a}$$

where w is random, the problem turns into a stochastic programming problem. The function $f(x, w)$ may be nonsmoth.

The special type of stochastic programming problems with recourse considered here are of the form

$$\min\ \{F(x) + Q(x)\}, \tag{3}$$

$$Q(x) = E\ \min\ \{q^+y^+ + q^-y^-; y^+ - y^- = h(w) - T(w)x,\\ 0 \leq\ y^+ \leq\ d^+,\quad 0 \leq\ y^- \leq\ d^-\}, \tag{4}$$

subject to linear constraints (5) similar to (2). For every value of w, the following equation

$$y_i^+(w) \cdot y_i^-(w) = 0, \tag{5}$$

must hold, i.e., for any w one of these two variables must equal 0.

The equalities in (4) determine the so-called stochastic constraints. The linear function in (4) determines the value of the recourse function. The vectors $q^- and q^+$ contain specific recourse coefficients.[1]

Each entry of the vector $h(w)$ or row of $T(w)$ may be deterministic, random with a discrete or standard uniform, normal and other distributions, or random with the distribution generated in the user's subroutine.

[1] Notice that the presence of upper bounds on y^+ and y^- implies that the considered problem is a generalization of the simple recourse one (editorial remark).

3. LINK BETWEEN DETERMINISTIC AND STOCHASTIC PROBLEMS

An important source of stochastic optimization problems is linear and nonlinear programming problems under constraints containing random values. As a rule, a deterministic problem is solved first, and then, after analyzing the obtained solution, its stochastic analog is created.

To analyze the solution of a stochastic problem it may be useful to compare its solution with that of a deterministic problem. For that purpose our software supports the linkage between deterministic and stochastic problems.

Let a LP problem have the form

$$\min cx \tag{6}$$

subject to

$$Ax = b, \quad x \geq 0. \tag{7}$$

Let some entries $a_{i,j}$ and some entries b_i be random values. Let us extract from (7) the constraints containing random values in the A and b rows. Let us form the matrix $T(w)$ and the vector $h(w)$ from these constraints, and the matrix A^1 and the vector b^1 from those remaining. So, A^1 and b^1 consist entirely of deterministic entries, whereas each $T(w)$ row or the corresponding $h(w)$ entry should have at least one random value. Then the constraints (7) may be expressed as follows:

$$A^1 x = b^1 \tag{8}$$

$$T(w)x = h(w). \tag{9}$$

The requirement for equality (9) to hold in every realization of w often leads to the absence of a feasible solution of problem.

Let us allow for discrepancies y^+, y^-

$$\begin{aligned} y_i^+(w) &= \max\,(h_i(w) - [T(w)x]_i, 0) \geq 0 \\ y_i^-(w) &= \max\,([T(w)x]_i - h_i(w), 0) \geq 0 \end{aligned} \tag{10}$$

express the discrepancies corresponding to the violation of the constraints (9) motivated by the uncertainty of the source data.

Using the well-known economic interpretation of LP problems ("input – output" models) let us consider the $a_{i,j}$ coefficients as the specific input, the b_i as the available resources, the $y^-(w)$ as discrepancies due to the shortage of resources, and the $y^+(w)$ as the discrepancies due to the surplus of resources. It is obvious that in every realization of w only one possibility is observed. Therefore for every w

$$y_i^+(w) \cdot y_i^-(w) = 0.$$

The violation of constraints leads to additional losses for their correction which may be considered proportional to the discrepancies. Let us specify the specific unit losses due to the shortage and the surplus of resources as q^- and q^+ . Then the total losses caused by the violation of constraints (9) are

$$q^- y^-(w) + q^+ y^+(w), \tag{11}$$

where

$$y^+(w) - y^-(w) = h(w) - T(w)x. \tag{12}$$

Minimizing the losses with respect to $y^+(w)$ and $y^-(w)$, we shall obtain the function $Q(x, w)$ expressing the minimum losses for every x.

Adding the function $Q(x) = E\, Q(x, w)$ to the objective function, we shall obtain a problem with recourse.

4. ALGORITHMS FOR SOLVING PROBLEMS

One of the multiplicative simplex methods is implemented for solution of linear programming problems.

For classical ("smooth") NLP problems one of the linearization methods described in [3] is implemented. For problems of nondifferentiable optimization so called r_α algorithm based on subgradient technique is used, see [4].

For stochastic problems two methods based on the calculation of the subgradient $\hat{F}_x(x)$ of the objective function $F(x) = E\, f(x, w)$ are used. $E\, f(x, w)$ is a multidimensional integral and generally its calculation is very difficult task. For the number of particular but very important cases calculations of subgradient $\hat{F}(x)$ can be carried out analyticaly and subgradient $\hat{F}_x(x)$ is the "usual" gradient $F_x(x)$. For example, it is true for two – stage problem (3) – (4) with linear objective function and for the number of standard distributions of $h(w)$, $T(w)$. For this problems classical methods of "smooth" nonlinear programming may be used.

The second method is based on the calculation of the stochastic quasigradient [2] of objective function (1a). In particular the subgradients $\hat{f}_x(x,w)$ for every w are estimates of the set $\hat{F}_x(x)$. The appropriate methods are following

$$x^{s+1} = \prod_x (x^s - \rho_s d^s), \quad s = 0,1\ldots \tag{13}$$

where ρ_s is the stepsize, d^s is the step direction, x^s is the solution vector estimate (current point) and Π_x is the projection on the set X.

The operation of projection represents a special quadratic programming problem. To solve it, a special modification of the conjugate gradients method is used. To form the direction vector d^{s+1}, the vector d^s and the stochastic quasigradient $g^s(x^s)$ are used. The calculation is executed as follows:

$$d^{s+1} = (g^s + \gamma_s d^s)/(1+\gamma_s), \quad d^0 = 0$$

where γ_s is a scalar factor. The condition:

$$< d^s, g^s > \geq 0$$

is verified in every iteration. If this condition is violated, then $\gamma_s = 0$. Calculation of r_s and γ_s is based upon [5], i. e.

$$r_s = r_0 exp(< \Delta^s, g^s > /Z^s),$$
$$\gamma_s = \gamma_0 exp(< \Delta^{s-1}, g^s > /Y^s),$$
$$\Delta^s = x^s - x^{s-1},$$

where r_0, γ_0 are constants and Z^s, Y^s are estimates of the $< \Delta^s, g^s >$ and $< \Delta^{s-1}, g^s >$ average values respectively. One more peculiarity of the algorithm is the method of calculating the stochastic quasigradient g^s for the objective function (3), i. e.

$$g^s = \hat{F}_x(x) + \hat{Q}_x(x,w),$$

where $\hat{F}_x(x)$ is an element from the subdifferential $\partial_x F(x)$ of convex function F(x) and $\hat{Q}_x(x,w)$ is an element from the subdifferential $\partial_x Q(x,w)$ with respect to the first component x.

The recourse function Q(x) is a separable one, i.e.

$$Q(x) = \sum_i E\,Q_i(x,w),$$

$$Q_i(x,w) = \min\ \{q_i^+ y_i^+ + q_i^- y_i^-; y_i^+ - y_i^- = h_i(w) - [T(w)x]_i \\ 0 \le y_i^+ \le d_i^+,\quad 0 \le y_i^- \le d_i^-\}$$

The stochastic quasigradient of every term of the separable function $Q(x)$ is calculated as

$$\nabla_x Q_i(x,w) = \begin{cases} -\quad q^+ T_i(w), & h_i(w) - [T_i(w)x] \ge 0 \\ \quad q_i^- T_i(w), & h_i(w) - [T_i(w)x] \le 0. \end{cases} \tag{14}$$

In some cases, if the distribution of $h_i(w)$ is known, we may calculate not only the stochastic quasigradient $\nabla_x Q_i(x,w)$, but also the subgradient $\hat{Q}_x(x)$

$$\hat{Q}_x(x) \in\ \partial_x Q_i(x) = E\,\partial_x Q_i(x,w).$$

In that way we may analytically calculate the subgradient for these $Q_i(x)$ through the corresponding formulae.

One more peculiarity of the algorithm is the technique of forming the set X on which the projection is performed. The feasible set may be represented as

$$K = K_1 \cap K_2,$$

$$K_1 = \left\{x : l \le \begin{pmatrix} x \\ Ax \end{pmatrix} \le u\right\},$$

$$K_2 = \{x:\ \forall w \in\ \Omega\, \exists 0 \le y^+ \le d^+, 0 \le y^- \le d^-, \\ y^+ - y^- = h(w) - T(w)x\}$$

Here Ω denotes a set of values accepted by the random value w. K_1 stands for a set of fixed constraints and K_2 is a set of induced constraints.

Using the induced constraints implies certain difficulties. First, the set of induced constraints is specified indirectly. Second, if a point does not belong to the set of induced constraints, then the stochastic quasi-gradient at this point does not exist. To overcome these difficulties the set X^s is formed in a special way at every iteration and the projection is executed on the set $X^s = K_1 K_{2,}^s$, where K_2^s is the linear approximation of the set of induced constraints in the vicinity of the point x^s. More precisely,

$$K_2^{s+1} = \begin{cases} K_2^s, & \text{if } \; y^+ \le d^+, \quad y^- \le d^- \\ K_2^s \cap \{x : h_i(w) - [T(w)x]_i \le d_i^+ \quad \forall i : y_i^+ > d_i^-\} \cap \\ \cap \{x : [T(w)x]_i - h_i(w) \le d_i^- \quad \forall i : y_i^- > d_i^-\}. \end{cases}$$

There is a special technique of forming the approximation set K_2^s which adds new constraints and automatically deletes inactive constraints.

5. INPUT AND OUTPUT DATA

The file structure depends on generalization of the international de facto MPS standard in the mathematical programming and reflects an attempt to ensure a certain compatibility with other systems for linear and nonlinear programming.

To use the software discussed it is not necessary to know the file structure, because all the input data are entered by means of specialized editors in matrix – graphic representation.

6. SOFTWARE FUNCTIONS

For solving problems, the following functions can be used:

- creating and modifying linear deterministic source data by means of the matrix editor;
- creating and modifying the user's subroutine for nonlinear components and for simulation of various distributions (optionally);
- creating and modifying the formula for nonlinear components by means a formular editor (optionally);
- solving LP and NLP (nondifferentiable) problems and performing post – optimum analysis;
- converting a deterministic problem to its stochastic analog and vice versa by means of the "convertor" subsystem;
- creating and modifying linear stochastic components for a stochastic problem by means of the stochastic matrix editor;
- solving stochastic optimization problems and performing post – optimum analysis.

In every algorithm iteration the current x^s point is written into the special file, which can be used later as initial point. During calculations the software displays the graph whose X axis represents the algorithm iteration numbers k = 1, 2, ..., and Y axis represents the values $F(x^k)$ or evaluations $F_k(x^k, w^k)$ of the objective function in these iterations. You may change the graph and to control the algorithm in interactive mode:

- change the current step value;
- clear the screen and redraw the graph with automatic rescaling on the Y axis;
- multiply the scale factor on the Y axis;
- divide the scale factor on the Y axix;
- quit the optimizer.

7. SUMMARY

Here the three software systems are under consideration.

1. Stochastic Nonlinear Optimization (SNLP) system [6] developed within the framework of the agreement between International Institute for Applied System Analysis (IIASA, Austria) and the Glushkov Institute of Cybernetics (USSR). SNLP system is designed to solve smooth NLP problems and their stochastic generalization (two stage stochastic problems) based on stochastic quasigradient technique.

2. Stochastic Linear Optimization (SLP) system developed in the Glushkov Institute of Cybernetics. SLP system is designed to solve LP problems and their stochastic generalization based on effective methods of smooth NLP.

3. Nondifferentiable and Stochastic Optimization (START) system developed in the Glushkov Institute of Cybernetics. START system is designed to solve nonlinear problems of nondifferentiable and stochastic optimization and the stochastic generalization of NLP problems based on stochastic quasigradient technique.

All of these systems are compatible one another.

REFERENCES

[1] Dantzig G. and Mandansky A., *On the Solution of Two - stage Linear Programs under Uncertainty*, Proceedings of the Fourth Berkeley Symposium on Mathematical Statistics and Probability, vol. 1, University California Press, Berkeley, 1961, pp. 165 - 176.

[2] Ermoliev Yu. and Wets R. B. (eds), *Numerical Techniques for Stochastic Optimization*, Springer - Verlag, Heidelberg, 1988.

[3] Pshenichni B., *Linearization Methods*, Nauka, Moscow, 1983. (Russian)

[4] Shor N. Z., *Generalized Gradient Method of Nondifferentiable Optimization Employing Space Dilation Operators*, Mathematical Programming: The State of the Art, Springer-Verlag, Berlin, 1983, pp. 501-529.

[5] Uryasiev S. P., *Stochastic Quasi-Gradient Algorithms with Adaptively Controlled Parameters*, IIASA Working paper WP-86-32 (1986), IIASA, Laxenburg, Austria.

[6] Roenko N., Loskutov V., Uryasiev S., *Stochastic Nonlinear Programming System*, Working paper WP-89-075 (1989), IIASA, Laxenburg, Austria.

UKRAINIAN ACADEMY OF SCIENCES, INSTITUTE OF CYBERNETICS, PROSPECT AC. V. M. GLUSHKOV, 252 207 KIEV, UKRAINA; TEL: (+7044) 266 55 59, TELEX: 131 272 GIC SU.

E-mail: CRIT%SOFTGIC.KIEV.UA.@RELAY.USSR.EU.NET

SOME COMPUTATIONAL ASPECTS OF GRAPHICAL MODEL SELECTION

DAVID EDWARDS

Novo Nordisk A/S, Denmark

ABSTRACT. Two alternative approaches to graphical model selection — stepwise edge elimination, and a so-called fast method proposed by Edwards and Havránek (1985, 1987) — are described and compared. Some emphasis is given to specific non-numerical computational aspects: in particular, an efficient algorithm for the dual representation problem is described. The model selection methods are applied to a contingency table concerning risk factors for coronary heart disease.

1. INTRODUCTION

The proper choice of statistical model is a key issue in applied statistics, and, indeed, in many of the empirical sciences. This paper sets out to examine several different approaches to the task based on one specific type of models, graphical models.

Graphical models are statistical models that can be represented by graphs in the mathematical sense of the word, that is, objects consisting of nodes and edges between the nodes. The nodes represent the variables in the model, and edges represent conditional dependences, in the sense that if an edge between two variables is absent then the variables are conditionally independent given the remaining variables. In this paper we consider models corresponding to undirected graphs only. We draw the nodes either as dots (for discrete variables) or as circles (for continuous variables).

Figure 1 shows a graphical model for three variables — age, sex and occupation. The missing edge between age and occupation indicates

---

1990 *Mathematics Subject Classification*. 62H15, 62H17.

Key words and phrases. Graphical models, loglinear models, fast model selection, backwards edge elimination, dual representation algorithm.

that under the model age and occupation are independent given sex, which we write using a standard notation (Dawid, 1979) as

$$\text{age} \perp \text{occupation} \mid \text{sex}.$$

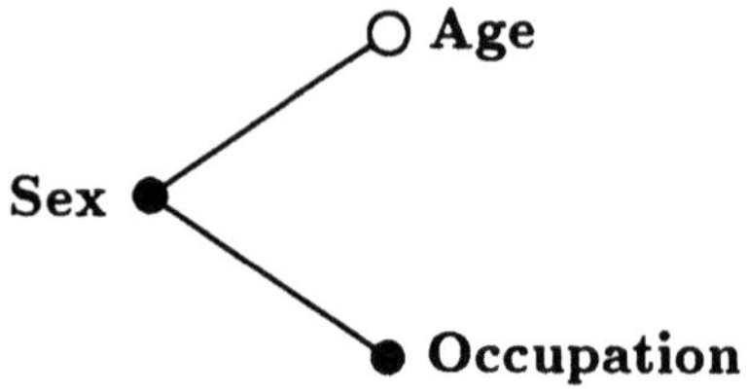

Figure 1.

In this context, model selection amounts to choice of graph: given a set of data, choose the simplest graph or graphs consistent with the data. This is the problem which will be treated below.

From a slightly broader perspective, the modelling process can be split into three stages: choice of model class, selection of a model from the class, and parameter estimation under the model. Formally it seems difficult, if not impossible, to distinguish between the three stages. Thus all three ought to be considered part of model selection in the proper sense.

Moreover most statistical theory regards the model as given: no account is taken of the uncertainty involved in model choice. This may perhaps account for the importance of model selection to statisticians: although in some contexts a formal statistical model is not of primary concern, the statistician feels forced to choose one in order to be able to apply the standard arsenal of statistical methods.

Most methods for model selection consist of:

(1) a model class;
(2) a criterion for judging the suitability of a model in the class, and;
(3) a search strategy for finding the most suitable model or models.

Section 2 below describes various model classes that can be used as the basis for graphical model selection, and sections 3 to 5 describe and compare various search strategies. The suitability criteria used are derived from significance tests.

We note briefly that the suitability criteria can be based on other frames of reference. For example, criteria derived from information theory can be used (Akaike, 1974). Greisser and Eddy (1979) describe a predictive approach. Recently there has been a resurgence of interest in Bayesian methods. Schwarz (1978) introduced a Bayesian equivalent of the AIC, the Bayesian Information Criterion. Raftery (1986) described the computation of Bayes factors for loglinear models with vague prior information. Dawid and Lauritzen (1989) provide general results concerning Bayesian inference for decomposable graphical models. Raftery and Madigan (1991) exploit these results in a selection procedure for contingency table models.

Furthermore, it is implicitly assumed here that all data are present, and that models are judged by their consistency with these data. In other contexts data may be available sequentially, so that model selection becomes a sequential process (Dawid, 1984). In probabilistic expert systems (Pearl, 1988, Lauritzen & Spiegelhalter, 1988), a topic closely related to graphical models, knowledge is usually acquired sequentially. The term knowledge acquisition is usually understood to mean parameter estimation for a fixed graph, but can also refer to what statisticians would call model selection, i.e. choice of graph.

2. Model Classes

In this section we describe specific model families that can be used for the type of model selection we are considering. We require model families for which there is a 1 − 1 correspondence between models and their independence graphs.

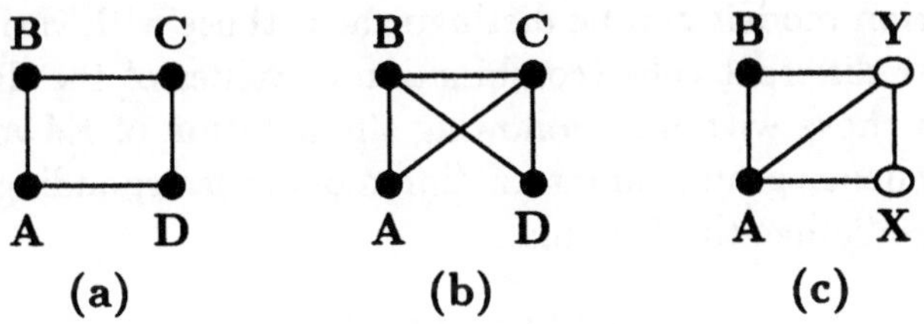

Figure 2.

For purely discrete data, one well-known possibility is the so-called graphical loglinear models introduced by Darroch, Lauritzen and Speed (1980). These constitute a subclass of loglinear models, whose non-zero interaction terms are given by the cliques (maximal complete subsets) of the independence graph. For example, the graphical models corresponding to Figures 2a and 2b have formulae

$$AB, BC, CD, AD$$

and

$$ABC, BCD,$$

respectively.

For sparse contingency tables, Whittaker (1990) advocates the use of first-order loglinear models, i.e. hierarchical loglinear models with at most two-factor interactions. The first-order models corresponding to Figures 2a and 2b have formulae

$$AB, BC, CD, AD$$

and

$$AB, AC, BC, BD, CD,$$

respectively. The variable sets in these model formulae are termed generators.

For purely continuous (normally distributed) data, covariance selection models (Dempster, 1972, Wermuth, 1976b, Whittaker, 1990) are available. These are analogous to first-order loglinear models, since the multivariate normal distribution has in effect only first order interaction terms.

For mixed continuous and discrete variables, the family of graphical interaction models was proposed by Lauritzen and Wermuth (1989), and a parametric extension to these (hierarchical interaction models) was described in Edwards (1990). More precisely, two classes of graphical interaction models can be distinguished: those with variance homogeneity over different cells (combination of values of the discrete variables), and those without. Following the notation of Edwards (1990), the homogeneous graphical interaction model corresponding to the graph of Figure 2c has the formula

$$AB/AX, ABY/XY,$$

whereas the heterogeneous model has the formula

$$AB/AX, ABY/AXY, ABY.$$

Clearly first-order mixed models could also be defined, and as with purely discrete data, these would be useful in connection with small data sets, where for example maximum likelihood estimates may not exist for the models with more parameters.

In the remainder of the section we study slightly more general subfamilies that are in 1 − 1 correspondence with graphs. For ease of notation, we consider discrete models only, but the results can be easily extended to the mixed case.

Let Δ be a set of discrete variables, and suppose that m is a hierarchical loglinear model for Δ. The generators of m correspond to the maximal interaction terms that are not set to zero. Recall the notion of the dual representation of m (Edwards and Havránek, 1985) that is, as the minimal interaction terms that are set to zero. For example, let

$$\Delta = \{A, B, C, D, E\}$$

and

$$m = ABC, BCD, CDE.$$

Then the dual representation of m is

$$[AD, AE, BE]^-,$$

since the model is characterized by setting precisely the two−factor interactions between AD, AE and BE to zero. Note, incidentally, that graphical models are characterized as model whose dual generators all have size two.

The class of hierarchical models forms a lattice, i.e. a partially ordered set with two binary operators, meet ($\wedge$) and join ($\vee$). The join of m_1 and m_2, written $m_1 \vee m_2$, is the smallest model containing both m_1 and m_2: the join operation can be performed by concatenating two model formulae (written in the standard representation) and then removing superfluous generators. Similarly $m_1 \wedge m_2$ is the largest hierarchical model included in both m_1 and m_2: the operation is performed by concatenating two model formulae written in the dual representation and then removing superfluous dual generators. The meet operation corresponds to the intersection of two hypotheses.

We define the operation of removing the edge AB from m as the operation

$$m := m \wedge [AB]^-.$$

Thus in algorithmic terms, the edge elimination operation takes three steps:

(1) Calculate the dual representation of m.
(2) Append AB to this and remove superfluous dual generators.
(3) Reconvert to the standard representation.

For example, to remove the edge BC from the model

$$ABC, BCD, CDE$$

we rewrite this as

$$[AD, AE, BE]^-,$$

append BC to this to obtain

$$[AD, AE, BE, BC]^-$$

and convert back to the standard representation which is

$$AB, AC, BD, CDE.$$

Now let h be any hierarchical loglinear model on Δ whose interaction graph is complete, so that h must include all two-factor interaction terms. Then for any given graph on Δ, defined through missing edges $e_1, e_2 \ldots e_p$, say, we can define a model $h \wedge [e_1, e_2 \ldots e_p]^-$. Clearly there is a $1-1$ correspondence between such models and graphs with vertex set Δ.

The converse operation, edge addition, can be performed as follows. Suppose we are given a model m to which we wish to add an edge e, say, and we also have given a model h as just described, so that we know that

$$m = h \wedge [e_1, e_2 \ldots e_p]^-$$

for some edges $e_1, e_2 \ldots e_p$. Then to add an edge e we first re-write m in its dual representation. Since e is not present in m, e will be one of the dual generators. Remove e to form the model m', say. Then the required model is $h \wedge m'$.

Efficient implementations of the edge elimination and addition operations rely heavily on the ability to transform between the standard and the dual model representations. An algorithm for calculating the dual representation was described in Edwards and Havránek (1985), but experience has shown this to be relatively inefficient: we now give a more efficient algorithm. The Pascal language is used.

```
Type
  node     = 'A'..'Z';
  gen      = set of node;
  nodelist = array[1..maxnovar] of node;
  genlist  = array[1..maxnogen] of gen;
```

```
Var
  Delta :  gen; {set of discrete variables}
Procedure DualRep(Var s:  genlist; Var k:  integer);
Var
  i, p, q:  integer;
  t:  genlist;
  v:  nodelist;

Procedure Expand(u, poss:  gen; i:integer);
Var j:  integer; w :  gen;
Begin
  while (i<=k) and (not (u<=s[i])) do i := i+1;
  if (i>k) then AddToList(u, t, q) else
  Begin
    w := poss-s[i];
    for j := 1 to p do if (v[j] in w) then
    begin
      poss := poss-[v[j]];
      expand(u+[v[j]], poss, i)
    End;
  End;
End;

Begin
  (* Set v[1..p] to nodes in Delta *)
  q := 0;
  Expand([], Delta, 1);
  k := q;
  for q:= 1 to k do s[q] := t[q];
End;
```

Some brief comments may be helpful. The algorithm shown is a type of recursive two-level algorithm that has been studied in depth in Rohl (1984). On entry, `DualRep` is passed the current model formula, s: the number of generators is passed as the integer k. On exit, these are replaced by the dual representation.

The procedure `Expand` is passed a current candidate for a dual generator, u, and a set of possible vertices with which u can be expanded. If u is not contained in any generator, it is added to the list of dual generators: otherwise, `Expand` is called recursively. In this way a depth-first search of generators is performed.

The procedure `AddToList` adds the generator u to the generator list t, where q is the length of the list. More precisely, it sets

$$t := \min\{t \cup u\}$$

since the dual generators are to be stored in minimal form. (Recall that the dual generators are the minimal sets not contained in any standard generator). Otherwise, the algorithm would not generally return the dual generators in their minimal form.

Details of the similar algorithm to transform the dual representation back to the standard representation are left to the diligent reader.

The computational complexity of the algorithm has yet to be studied in detail, though some simulation studies performed by J.-H. Badsberg (personal communication) suggest that it may be exponential.

The algorithm is also useful in another context, described below in section 5.

3. Stepwise Edge Elimination

Probably the most commonly used approach to graphical model selection is a stepwise one, in which edges are sequentially eliminated from the independence graph using asymptotic tests based on the deviance. A discussion of this approach is given in Edwards and Kreiner (1983).

The deviance difference between two models $\mathcal{M}_0 \subseteq \mathcal{M}_1$ is given as

$$\hat{d} = -2\ln\frac{\hat{\mathcal{L}}(\mathcal{M}_1)}{\hat{\mathcal{L}}(\mathcal{M}_0)} = 2\sum_i n(i)\frac{m_0(i)}{m_1(i)}$$

where $\hat{\mathcal{L}}(\mathcal{M}_0)$ and $\hat{\mathcal{L}}(\mathcal{M}_1)$ are the maximised likelihoods under $\mathcal{M}_0$ and $\mathcal{M}_1$, and $n(i)$, $\hat{m}_0(i)$ and $\hat{m}_1(i)$ are respectively the observed count, the fitted count under $\mathcal{M}_0$ and the fitted count under $\mathcal{M}_1$ for cell i. The sum is over all cells in the table.

It may be shown that $\hat{d}$ is asymtotically χ^2-distributed, with degrees of freedom equal to the difference in number of free parameters between $\mathcal{M}_0$ and $\mathcal{M}_1$.

We will not discuss methods for performing maximum likelihood estimation in the present paper, other than by remarking that two general approaches can be adopted. The iterative proportional scaling (IPS) algorithm (Fienberg, 1970) is simple to program, is insensitive to choice of starting values, and has modest memory requirements. In contrast, a Gauss-Newton algorithm, though it does not have these assets, has the advantage that it provides estimates of asymptotic covariances for the parameter estimates.

As an example of stepwise edge deletion we consider analysis of the data shown in Table 1, that treat risk factors for coronary heart disease. For sample of 1841 car-workers, the following information was recorded: A (smoker: Y/N), B (strenuous mental work: Y/N), C (strenuous physical work: Y/N), D (systolic blood pressure), E (ratio of β to α lipoproteins), and F (family history of coronary heart disease).

The data form a six-way contingency table.

			B	no	no	yes	yes
			A	no	yes	no	yes
F	E	D	C				
negative	<3	<140	no	44	40	112	67
			yes	129	145	12	23
		>=140	no	35	12	80	33
			yes	109	67	7	9
	>=3	<140	no	23	32	70	66
			yes	50	80	7	13
		>=140	no	24	25	73	57
			yes	51	63	7	16
positive	<3	<140	no	5	7	21	9
			yes	9	17	1	4
		>=140	no	4	3	11	8
			yes	14	17	5	2
	>=3	<140	no	7	3	14	14
			yes	9	16	2	3
		>=140	no	4	0	13	11
			yes	5	14	4	4

Table 1.

Table 2 shows the progress of the stepwise edge elimination procedure applied to the data in Table 1.

Starting with the full (saturated) model, the elimination of each of the remaining eligible edges is examined at each step. Nested LR tests at fixed level $\alpha = 0.05$ are performed. Table 2 shows the value of the LR statistic, the associated degrees of freedom and the $p-$value. The "least significant" edge, i.e. the edge with the largest $p-$value $> \alpha$, is removed at each step. (Such edges are marked with "-"s in Table 2). In addition the principle that if a model is rejected, then all its submodels should also be rejected, is applied. Here the principle means that if elimination of an edge is rejected at one step (i.e. the associated $p-$value is less than α), the edge is not eligible for elimination at subsequent steps. (These edges are marked with "+"s in Table 2).

The process continues until no further edges can be removed.

The procedure selects the model

$$ABC, BF, ACE, ADE.$$

Its graph is shown in Figure 3a.

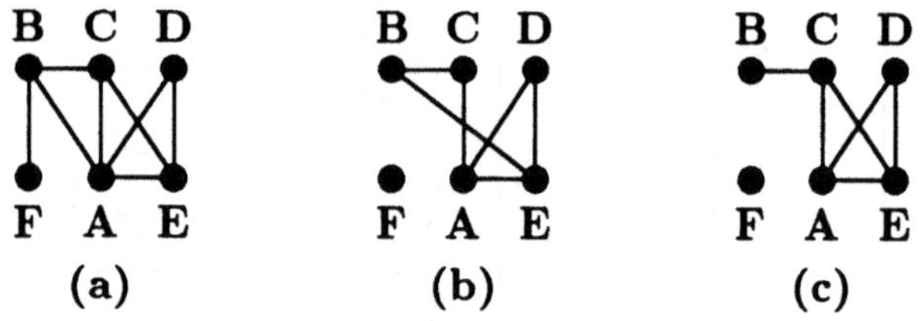

Figure 3.

The model has deviance 51.3587 on 46 df ($P = 0.2718$):

Consider the tests for elimination of a specific edge, e.g. EC, in Table 2. These test the hypothesis $E \perp C \mid ABDF$ under changing models. As the models get progressively simpler, so the power of the test increases: in other words, the initial tests have relatively low power. Thus the risk of Type II error (false elimination of edges) is highest in the initial columns of Table 2.

A variant of the procedure designed to counteract this is the following. At each step examine the edge elimination tests with the largest power (least degrees of freedom) and remove the least significant edge

Edge	Step 1				Step 2			
[BA]	22.651	16	0.123		15.574	8	0.048	+
[CA]	42.803	16	0.000	+				
[CB]	684.989	16	0.000	+				
[DA]	28.724	16	0.025	+				
[DB]	12.225	16	0.728	−				
[DC]	14.808	16	0.538		7.148	8	0.520	−
[EA]	40.024	16	0.001	+				
[EB]	17.226	16	0.371		11.363	8	0.181	
[EC]	18.629	16	0.288		15.548	12	0.212	
[ED]	31.059	16	0.013	+				
[FA]	21.305	16	0.167		16.531	12	0.168	
[FB]	22.787	16	0.119		15.743	8	0.046	+
[FC]	22.152	16	0.138		17.026	12	0.148	
[FD]	18.345	16	0.304		11.301	8	0.185	
[FE]	18.316	16	0.305		14.629	12	0.262	

Table 2.

Edge	Step 3				Step 4			
[BA]								
[CA]								
[CB]								
[DA]								
[DB]								
[DC]								
[EA]								
[EB]	11.363	8	0.181		6.738	4	0.150	
[EC]	12.735	8	0.121		7.547	4	0.109	
[ED]								
[FA]	15.318	10	0.120		8.585	5	0.126	
[FB]								
[FC]	12.237	8	0.140		7.049	4	0.133	
[FD]	6.367	4	0.173		2.290	2	0.318	−
[FE]	13.247	10	0.210	−				

Table 2. (continuation)

Edge	Step 5				Step 6			
[BA]								
[CA]								
[CB]								
[DA]								
[DB]								
[DC]								
[EA]								
[EB]	6.770	4	0.148	–				
[EC]	7.516	4	0.111		21.675	2	0.000	+
[ED]								
[FA]	7.385	4	0.116		7.385	4	0.116	
[FB]								
[FC]	7.018	4	0.134		7.018	4	0.134	–
[FD]								
[FE]								

Table 2. (continuation)

Edge	Step 7			
[BA]				
[CA]				
[CB]				
[DA]				
[DB]				
[DC]				
[EA]				
[EB]				
[EC]				
[ED]				
[FA]	2.658	2	0.264	–
[FB]				
[FC]				
[FD]				
[FE]				

Table 2. (continuation)

among these. In other words, instead of deciding which edge to remove from the p-values alone, sort edge by degrees of freedom (ascending) and then within these by p-values (descending). The intention of this approach would be to increase power (reduce Type II errors): it would be interesting to study whether or not this is in fact achieved.

4. Decomposable Stepwise Edge Deletion

Loglinear models with explicit maximum likelihood estimates, the so-called decomposable models (Goodman, 1970) have been studied by many authors. Wermuth (1976a) proposed using them as a basis for model selection, and Darroch, Lauritzen and Speed (1980) characterized them as graphical models with triangulated independence graphs, i.e. as having no cycles of length ≥ 4 without a chord.

Triangulated graphs have also been the subject of much study (see for example, Golumbik, 1980) and linear-time algorithms have been developed for recognizing the triangulated property (Tarjan and Yannakakis, 1984: see also Tarjan, 1985).

Decomposability of mixed graphical interaction models has been studied in depth in Lauritzen and Wermuth (1989), Frydenberg and Lauritzen (1989) and Frydenberg (1990): for the corresponding graph-theoretic results see Leimer (1989).

It is attractive to restrict model selection to this subclass for a variety of reasons. Firstly, computational efficiency: the ability to use explicit formulae for ML estimates can reduce computational time substantially. This is particularly important for high-dimensional problems. Secondly, for more theoretical reasons. For example, Bartlett corrections for decomposable models have been derived, both in the discrete case, i.e. loglinear models (Williams, 1976) and in the continuous case, i.e. covariance selection models (Porteous, 1985). Moreover, exact conditional tests for nested decomposable models differing by one edge are readily computed: for contingency tables, these correspond to tests for conditional independence in 3-way tables, in the covariance selection context, these will be F-tests, and for mixed models, both forms may occur (Frydenberg and Lauritzen, 1989). Similarly, exact overall goodness-of-fit tests for decomposable loglinear models can be calculated (Badsberg, 1991).

Liberation from dubious asymptotic is particularly compelling in the analysis of categorical data, since high-dimensional contingency tables are almost invariably sparse. Moreover, the computations required for the exact tests, although intensive, are easily generalizable, in the sense that the conditional distributions of non-standard statistics are easily calculated. So, for example, exact tests can be based on measures of conditional association suitable to ordinal variables: this can give a substantial increase in power.

The program DIGRAM (Kreiner, 1989) adopts this approach to graphical model selection: see also Kreiner (1987). Monte Carlo methods enable the exact p-values to be calculated to any desired precision.

From an algorithmic point of view, the stepwise edge elimination procedure is easily restricted to decomposable models only, when the following result is applied. Starting with a decomposable model, a new model formed by removing an edge will be decomposable if and only if the edge occurs in one generator only (Edwards, 1984). For example, consider the decomposable model

$$ABCD, ABE, BCE.$$

The set $\{A, B\}$ occurs in more than one generator: consequently, the edge AB is not eligible for elimination, since this would lead to a non-decomposable model.

Table 3 shows the result of performing the restricted procedure to the data of Table 1. The same model as before is selected, after performing slightly fewer tests. The p-values shown are asymptotic. Exact tests could have been applied here, using either DIGRAM (Kreiner, 1989) or CoCo (Badsberg, 1991).

5. The EH-procedure

In this section we examine a different approach to model selection, proposed by Edwards and Havránek (1985, 1987), and which we here call the EH-procedure. Let $\mathcal{F}$ be the model family under consideration, and suppose that ϕ is a goodness-of-fit test for models $m \in \mathcal{F}$, i.e.

$$\phi_m = \begin{cases} 0 & \text{if } m \text{ is rejected} \\ 1 & \text{otherwise} \end{cases}$$

for each $m \in \mathcal{F}$. We denote model inclusion by $\prec$. If $m_1 \preceq m_2 \Rightarrow \phi_{m_1} \leq \phi_{m_2}$ we call ϕ coherent. If this is usually true, we call ϕ mildly noncoherent. In the contingency table context, ϕ_m will usually be based on the asymptotic likelihood ratio test against the full model, applied at fixed level α: we regard this test as mildly noncoherent.

Edge	Step 1				Step 2			
[BA]	22.651	16	0.123		15.574	8	0.048	+
[CA]	42.803	16	0.000	+				
[CB]	684.989	16	0.000	+				
[DA]	28.724	16	0.025	+				
[DB]	12.225	16	0.728	-				
[DC]	14.808	16	0.538		7.148	8	0.520	-
[EA]	40.024	16	0.001	+				
[EB]	17.226	16	0.371		11.363	8	0.181	
[EC]	18.629	16	0.288					
[ED]	31.059	16	0.013	+				
[FA]	21.305	16	0.167					
[FB]	22.787	16	0.119		15.743	8	0.046	+
[FC]	22.152	16	0.138					
[FD]	18.345	16	0.304		11.301	8	0.185	
[FE]	18.316	16	0.305					

Table 3.

Edge	Step 3				Step 4			
[BA]								
[CA]								
[CB]								
[DA]								
[DB]								
[DC]								
[EA]								
[EB]	11.363	8	0.181	-				
[EC]	12.735	8	0.121		22.530	4	0.000	+
[ED]								
[FA]								
[FB]								
[FC]	12.237	8	0.140					
[FD]	6.367	4	0.173		6.367	4	0.173	-
[FE]								

Table 3. (continuation)

Edge	Step 5				Step 6			
[BA]								
[CA]								
[CB]								
[DA]								
[DB]								
[DC]								
[EA]								
[EB]								
[EC]								
[ED]								
[FA]					7.385	4	0.116	
[FB]								
[FC]					7.018	4	0.134	-
[FD]								
[FE]	4.576	4	0.333	-				

Table 3. (continuation)

Edge	Step 7			
[BA]				
[CA]				
[CB]				
[DA]				
[DB]				
[DC]				
[EA]				
[EB]				
[EC]				
[ED]				
[FA]	2.658	2	0.264	-
[FB]				
[FC]				
[FD]				
[FE]				

Table 3. (continuation)

For convenience we say a model is accepted instead of the more correct "non-rejected". The approach seeks to identify the simplest models in $\mathcal{F}$ that are accepted, and is based on two rules:

(i) If a model m is rejected, then all submodels of m are inferred to be rejected. (We call them w-rejected).

(ii) If a model m is accepted, then all models that include it are inferred to be accepted. (We call them w-accepted).

The EH-procedure is based on two model sets, $\mathcal{A}$ and $\mathcal{R}$. The models fitted during the procedure are tested and placed in $\mathcal{A}$ if they are accepted or in $\mathcal{R}$ if they are rejected. At any stage, based on the current contents of $\mathcal{A}$ and $\mathcal{R}$, we calculate two new model sets: the r-dual of $\mathcal{A}$, defined as

$$D_r(\mathcal{A}) = \max\{m \in \mathcal{F} \mid a \not\preceq m, \forall a \in \mathcal{A}\}$$

and the a-dual of $\mathcal{R}$, defined as

$$D_a(\mathcal{R}) = \min\{m \in \mathcal{F} \mid m \not\preceq r, \forall r \in \mathcal{R}\}.$$

Figure 4 illustrates how the sets $\mathcal{A}$ and $\mathcal{R}$ divide $\mathcal{F}$ into three disjoint subsets: the w-accepted models, the w-rejected models, and the remainder, say $\mathcal{U}$, i.e. the models that are neither w-accepted or w-rejected. It is easy to see that $\min(\mathcal{U}) = D_a(\mathcal{R}) \setminus \mathcal{A}$ and $\max(\mathcal{U}) = D_r(\mathcal{A}) \setminus \mathcal{R}$.

The next step of the procedure consists of testing the models in either $\min(\mathcal{U})$ or $\max(\mathcal{U})$. Normally the set with the smallest number of models is tested, but other choices are possible. The procedure continues until $\mathcal{U}$ is empty, at which point $A = D_a(\mathcal{R})$ or equivalently $R = D_r(\mathcal{A})$. The procedure can either start from scratch ($\mathcal{A} = \mathcal{R} = \emptyset$, so that $D_a(\mathcal{R}) = \min(\mathcal{F})$ and $D_r(\mathcal{A}) = \max(\mathcal{F})$, or by fitting some initial set of models. This initial set can be chosen arbitrarily, but the assignment to $\mathcal{A}$ and $\mathcal{R}$ must be coherent, i.e. $m_1 \in \mathcal{A}$, $m_2 \in \mathcal{R}$, $m_1 \preceq m_2$ must not hold.

If the test used is coherent, the sets A and R will be unique, i.e. the same set-pairs will be found irrespective of which initial set and which choices ($\min(\mathcal{U})$ or ($\max(\mathcal{U})$) are made. If the test is mildly noncoherent, there will be what might be described as a fuzzy boundary between the w-accepted and w-rejected regions, in which the various solution set-pairs $(\mathcal{A}, \mathcal{R})$ are constrained to lie.

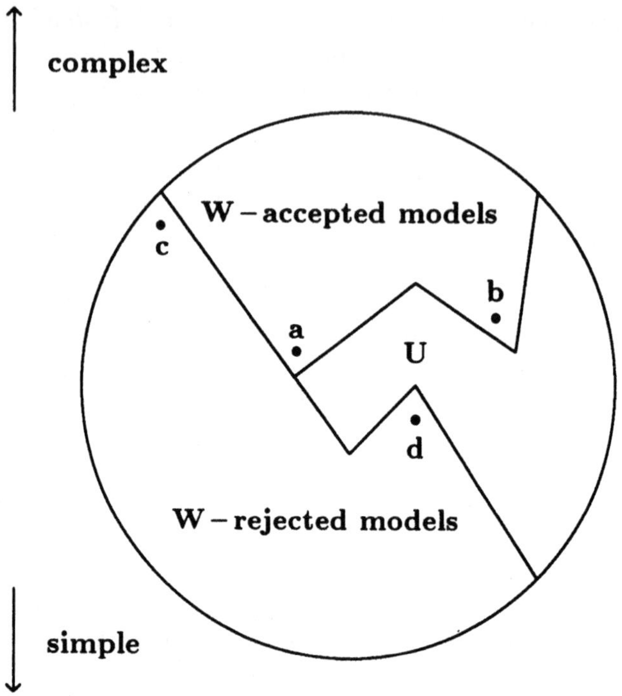

Figure 4.

We now briefly describe how the EH – procedure can be modified to perform restricted search to subfamilies of $\mathcal{F}$ of the form

(1) $\{m \in \mathcal{F} \mid m \preceq m_1\}$,
(2) $\{m \in \mathcal{F} \mid m_0 \preceq m\}$, or
(3) $\{m \in \mathcal{F} \mid m_0 \preceq m \preceq m_1\}$.

The modification is very simple, consisting of initially assigning models (without fitting) to $\mathcal{A}$ or $\mathcal{R}$. The first two cases were sketched in Edwards and Havránek (1987). By initially assigning $D_r(\{m_1\})$ to $\mathcal{A}$, subsequent search is restricted to $\{m \in \mathcal{F} \mid m \preceq m_1\}$. Similarly, by initially assigning $D_a(\{m_0\})$ to $\mathcal{R}$, subsequent search is restricted to $\{m \in \mathcal{F} \mid m_0 \preceq m\}$.

However, combining these restrictions to a set of the form $\{m \in \mathcal{F} \mid m_0 \preceq m \preceq m_1\}$ is trickier. Simply assigning $D_r(\{m_1\})$ to $\mathcal{A}$ and $D_a(\{m_0\})$ to $\mathcal{R}$ may well violate the coherence requirement, which would cause the procedure to fail. To enforce coherence we can assign $D_r(\{m_1\})$ to $\mathcal{A}$ and $D_a(\{m_0\} \cup D_r(\{m_1\}))$ to $\mathcal{R}$. If $\mathcal{F}$ is a lattice, then using Lemma 3 of Edwards and Havránek (1987) we can simplify

$$D_a(\{m_0\} \cup D_r(\{m_1\}) = \min\{m \wedge m_1 \mid m \in D_a(\{m_0\}\}$$

which, with a certain license, we can abbreviate to $D_a(\{m_0\}) \wedge m_1$. Equivalently we could set $D_a(\{m_0\})$ to $\mathcal{R}$ and $D_r(\{m_1\}) \wedge m_0$ to $\mathcal{A}$.

Thus in the graphical modelling context we can restrict to model sets of the form $\{m \in \mathcal{F} \mid m_0 \preceq m \preceq m_1\}$ provided that m_0 is a submodel of m_1 defined through conditional independence restrictions only, i.e. that $m_0 = m_1 \wedge [e_1, \ldots, e_p]^-$ for some edges $e_1, \ldots, e_p$.

We now turn to other computational aspects of the procedure. The main computational burden consists of two tasks: fitting the models, and calculating the duals, $D_r(\mathcal{A})$ and $D_a(\mathcal{R})$. We do not discuss the first of these here: in general the iterative proportional scaling algorithm (Fienberg, 1970) must be used. For large dimensional problems the computation of the dual can involve considerable calculation.

For models specified as graphs, or more generally for any model family with a binary lattice structure, calculation of the duals is formally equivalent to the dual representation problem. Here edges are equated with variables (letters), models (edge sets) with generators, and model lists with generator lists (model formulae). The dimension of the problem increases from p (the number of variables) to $p(p-1)/2$ (the number of edges). The algorithm sketched above in section 2 can be used: however note that in the model selection context, more sophisticated data representations will be appropriate. For example, model lists are better stored as linked lists rather than fixed length arrays.

The EH – procedure is implemented in the programs MIM (Edwards, unpublished manuscript) and CoCo (Badsberg, 1991).

The application of the EH – procedure to the data in Table 1 was reported in detail in Edwards and Havránek (1985). Using $\alpha = 0.05$, and without specifying an initial set of models, the procedure selected two models:

$$AC, ADE, BC, BE, F,$$

which has a deviance of 63.01 on 50 df ($P = 0.1023$), and

$$ACE, ADE, BC, F$$

which has a deviance of 62.08 on 49 df ($P = 0.0994$). The models are shown in Figures 3b and 3c.

Finally, we consider restricting the EH – procedure to decomposable models only. Let $\mathcal{D}$ and $\mathcal{G}$ be the classes of decomposable, respectively graphical models for some fixed set of vertices. We have seen above how the procedure can be applied to find the simplest models in $\mathcal{G}$ consistent with the data. To restrict to decomposable models only, a naive approach would calculate

$$D_r^+(\mathcal{A}) = D_r(\mathcal{A}) \cap \mathcal{D}$$

$$D_a^+(\mathcal{R}) = D_a(\mathcal{R}) \cap \mathcal{D}$$

where $D_r(\mathcal{A})$ and $D_a(\mathcal{R})$ are calculated as before, and then fit at each stage either $D_r^+(\mathcal{A}) \setminus \mathcal{R}$ or $D_a^+(\mathcal{R}) \setminus \mathcal{A}$, until both of these sets are empty. This would ensure that only decomposable models are fitted. However, there is no assurance that the status of all decomposable models is determined, i.e. that $\mathcal{U} \cap \mathcal{D} = \emptyset$.

Instead, a better approach is to calculate directly

$$D_r^d(\mathcal{A}) = \max\{m \in \mathcal{D} \mid a \not\leq m, \forall a \in \mathcal{A}\}$$

and

$$D_a^d(\mathcal{R}) = \min\{m \in \mathcal{D} \mid m \not\leq r, \forall r \in \mathcal{R}\}.$$

The EH – procedure can then be applied in the usual way, even though $\mathcal{D}$ is not a lattice (see Edwards and Havránek, 1987).

The algorithm sketched in Section 2 can easily be modified to calculate $D_r^d(\mathcal{A})$ by replacing the condition in line 3 in Expand, i.e.

$$\text{if } (i > k)$$

with the condition

$$\text{if } (i > k) \text{ and Decomp(u).}$$

Here u represents a model, as explained above, and Decomp is a Boolean function testing u for decomposability. A linear – time algorithm should be used for this: see Tarjan and Yannakakis (1984) for the pure case, or Leimer (1989) for the mixed case.

6. Comparison of the Selection Procedures

Both approaches described above are essentially heuristic methods which lead to the choice of simple models consistent with the data. The stepwise approach is derived by analogy with the well-known methods of stepwise regression analysis, and experience shows that it is an efficient method for choosing a single plausible model.

In contrast, the EH-procedure typically selects multiple models consistent with the data: it may be argued that this faithfully reflects inherent ambiguity, and therefore is an advantage. Since the tests used are based on overall goodness-of-fit rather than nested tests, for a fixed level α, the models chosen will typically be simpler than those selected by stepwise methods. Compare for example the models selected by the two approaches above.

The sampling properties of the stepwise procedure are complex, since it involves multiple tests. It can be shown that for a *fixed* sequence if decomposable models differing by one edge, the nested LR tests are mutually independent. However this does not seem to be very helpful, since obviously during stepwise selection the sequence of edges removed depends on the LR tests.

Rather more can be said about the sampling properties of the EH-procedure. Firstly, if a coherent test with a fixed level α is used, then we know (Edwards and Havránek, 1987) that the "true" model is contained in the set of w-accepted models with probability $\geq 1-\alpha$, so this set can be regarded as a $1-\alpha$ confidence set for the true model. If the test is only mildly coherent, then this holds only approximately.

Secondly, we can relate the procedure to the results presented elsewhere in this volume on multiple test procedures (Hommel and Bernhard, 1992). Consider the upwards-only version of the procedure, which takes $\min(\mathcal{F})$ as the initial set of models to be fitted, and at each stage fits $\min(\mathcal{U})$, i.e. $D_a(\mathcal{R})\setminus\mathcal{A}$. In the graphical case, this would start by fitting models with one edge, then models with two edges that both were rejected at the previous step, and so on. As noted by Havránek (1988) (see also Smith, 1990), this procedure is closed in the sense of Marcus, Peritz and Gabriel (1976), and thus controls the family wise error rate (FWE), i.e.

$$\Pr(\#\text{false rejections} > 0) < \alpha.$$

It follows that if a coherent test is used, the general EH-procedure also controls the FWE (since in this case the same models are selected).

If ϕ_m is not coherent, then the FWE is not controlled. If ϕ_m is mildly noncoherent, a common sense approach to controlling the FWE is to try to remove edges from each of the selected models. In other words, after the procedure has selected a set $\mathcal{A}$, test all models formed by removing an edge from each of the model in $\mathcal{A}$. In effect, this examines whether the "fuzzy boundary" extends below the selected models.

Although these results are interesting, the reader may well wonder whether they are sufficient: after all, the FWE can always be controlled just by accepting all models in $\mathcal{F}$. Classical test theory involves a balance between Type I and Type II error: in the present context controlling the FWE alone is analogous to controlling Type I error alone. This would suggest that a concept analogous to power seems needed, before a satisfactory basis for the comparative study of model selection procedures can be developed.

8. Acknowledgements

Helpful conversations with Svend Kreiner, Steffen Lauritzen, Jens-Henrik Badsberg, Morten Frydenberg and Poul Svante Eriksen are gratefully acknowledged. Particular thanks are due to Jens-Henrik Badsberg for help with the dual representation algorithm.

References

[1] Akaike H., *A New Look at the Statistical Model Identification*, IEEE AC **19** (1974), 716-23.

[2] Badsberg J. H., *A Guide to CoCo*, Research Report R91-43, Institute for Electronic Systems, Aalborg University, Aalborg, Denmark, 1991.

[3] Darroch J. N., Lauritzen S. L. and Speed T. P., *Markov Fields and Log Linear Interaction Models for Contingency Tables*, Ann. Statist. **8** (1980), 522-539.

[4] Dawid A. P., *Conditional Independence in Statistical Theory (with discussion)*, JRSS B **41** (1979), 1-31.

[5] Dawid A. P., *Statistical Theory: the Prequential Approach*, JRSS A **147** (1984), 278-292.

[6] Dawid A. P. and Lauritzen S. L., *Markov distributions, hyper-Markov laws and meta-Markov models on decomposable graphs, with application to Bayesian learning in expert systems*, BAYES report BR-10, 1989.

[7] Dempster A. P., *Covariance Selection*, Biometrics **28** (1972), 157-175.

[8] Edwards D., *Hierarchical Interaction Models*, JRSS B **52** (1990), 3-20.

[9] Edwards D. and Kreiner S., *The Analysis of Contingency Tables by Graphical Models*, Biometrika **70** (1983), 553-565.

[10] Edwards D., *A Computer Intensive Approach to the Analysis of Sparse Multidimensional Contingency Tables*, COMPSTAT'84 : Proceedings in Computational Statistics (Havránek T., Šidák Z. and Novák M., eds.), Physica-Verlag, Vienna, 1984, pp. 355-360.

[11] Edwards D. and Havránek T., *A Fast Procedure for Model Search in Multidimensional Contingency Tables*, Biometrika **72** (1985), 339 – 351.

[12] Edwards D. and Havránek T., *A Fast Model Selection Procedure for Large Families of Models*, JASA **82** (1987), 205 – 213.

[13] Edwards D., *A Guide to MIM 2.0.* (unpublished manuscript)

[14] Fienberg S. E., *An Iterative Procedure for Estimation in Contingency Tables*, Ann. Math. Statist. **41** (1970), 907 – 917.

[15] Frydenberg M., *Marginalization and Collapsibility in Graphical Interaction Models*, Ann. Statist. **18** (1990), 790 – 805.

[16] Frydenberg M. and Lauritzen S. L., *Decomposition of Maximum Likelihood in Mixed Graphical Interaction Models*, Biometrika **76** (1989), 539 – 555.

[17] Geisser S. and Eddy W. F., *A Predictive Approach to Model Selection*, JASA **74** (1979), 153 – 160.

[18] Golumbik M.C., *Algorithmic Graph Theory and Perfect Graphs*, Academic Press, New York, 1980.

[19] Goodman L., *The Multivariate Analysis of Qualitative Data: Interactions among Multiple Classifications*, JASA **65** (1970), 226 – 256.

[20] Havránek T., *A Procedure for Model Search in Multidimensional Contingency Tables*, Biometrics **40** (1984), 95 – 100.

[21] Havránek T., *Model Search in Large Model Families*, Proceedings of the First World Congress of the Bernoulli Society (Prokhorov Yu. A. and Sazonov V. V., eds.), vol. 2, VNU Sci. Press, Utrecht, 1987, pp. 767 – 778.

[22] Havránek T., *On Application of Statistical Model Search Techniques in Constructing a Probabilistic Knowledge Base*, Transactions of Eleventh Prague Conf. on Info. Theory, Stat. Dec. Funct. and Random Proc., Academia, Praha. (Will be published.)

[23] Hommel G. and Bernhard G.. (This volume)

[24] Kreiner S., *Analysis of Multi-Dimensional Contingency Tables by Exact Conditional Tests: Techniques and Strategies*, Scand. J. Statist. **14** (1987), 97 – 112.

[25] Kreiner S., *A Users Guide to DIGRAM - a program for discrete graphical modelling*, Research Report 89/10, Statistical Research Unit, University of Copenhagen, Denmark, 1989.

[26] Lauritzen S. L. and Spiegelhalter D. J., *Local Computations with Probabilities on Graphical Structures and their Application to Expert Systems (with discussion)*, JRSS B **50** (1988), 157 – 224.

[27] Lauritzen S. L. and Wermuth N., *Graphical Models for Associations Between Variables, some of which are Qualitative and some Quantitative*, Ann. Statist. **17** (1989), 31 – 57.

[28] Leimer H. G., *Triangulated Graphs with Marked Vertices*, Ann. Disc. Math. **41** (1989), 311 – 324.

[29] Marcus R., Peritz E. and Gabriel K. R., *On Closed Testing Procedures with Special Reference to Ordered Analysis of Variance*, Biometrika **63** (1976), 655 – 660.

[30] Morrison D. F., *Multivariate Statistical Methods*, McGraw – Hill, New York, 1967.

[31] Pearl J., *Probabilistic Inference in Intelligent Systems*, Morgan Kaufmann, San Mateo, 1988.

[32] Porteous B. T., *Improved Likelihood Ratio Statistics for Covariance Selection Models*, Biometrika **72** (1985), 474 – 475.

[33] Raftery A. E., *A Note on Bayes Factors for log – Linear Contingency Table Models with Vague Prior Information*, JRSS B **48** (1986), 249 – 250.

[34] Raftery A. E. and Madigan, *Model selection and accounting for model uncertainty in graphical models using Occams razor*, Technical report no. 213, University of Washington, Department of Statistics, Seattle, 1991.

[35] Reiniš Z., Pokorný J., Basika V., Tišerová J., Goričan K., Horáková D., Stuchlíková E., Havránek T. and Hrabovský F., *Prognostický význam rizikového profilu v prevenci ischemické choroby srdce*, Bratislavské Lékařské Listy **76** (1981), 137 – 150. (in Czech)

[36] Rohl J. S., *Recursion via Pascal*, Cambridge University Press, Cambridge, 1984.

[37] Schwarz G., *Estimating the Dimension of a Model*, Ann. Statist. **6** (1978), 461 – 464.

[38] Smith P., Ph.D. thesis (unpublished), Lancaster University, Lancaster, U.K.

[39] Tarjan R. E. and Yannakakis M., *Simple Linear – Time Algorithms to Test Chordality of Graphs, Test Acyclicity of Hypergraphs, and Selectively Reduce Acyclic Hypergraphs*, SIAM J. Comp. **13** (1984), 566 – 579.

[40] Tarjan R. E., *Decomposition by Clique Separators*, Discrete Maths **55** (1985), 221 – 232.

[41] Wermuth N., *Model Search in Multiplicative Models*, Biometrics **32** (1976 a), 253 – 263.

[42] Wermuth N., *Analogies between Muliplicative Models in Contingency Tables and Covariance Selection*, Biometrics **32** (1976 b), 95 – 108.

[43] Whittaker J., *Graphical Models in Applied Statistics*, Wiley, Chichester, 1990.

[44] Williams D. A., *Improved Likelihood Ratio Tests for Complete Contingency Tables*, Biometrika **63** (1976), 33 – 37.

NOVO NORDISK A/S, CNS DIVISION, NOVO NORDISK PARK, DK – 2760 MÅLØV, DENMARK; TEL: (+45 4446) 8888, FAX: (+45 4446) 3939.

MULTIPLE HYPOTHESES TESTING

GERHARD HOMMEL, GUDRUN BERNHARD

University of Mainz, Mainz

ABSTRACT. The paper is mainly concerned with multiple testing procedures which control a given multiple level α. General concepts for this purpose are the closure test and a modification which is independent of the special structure of hypotheses and tests. We consider improvements of this modification using information about the logical dependences (redundancies) within the system of hypotheses and present an efficient algorithm. Finally, we discuss some problems which are specific for hierarchical systems of hypotheses, e.g. in model search.

I. INTRODUCTION

I.1 The problem of multiple testing.

The problem of multiple hypotheses testing arises, if more than one statistical test has to be performed within the same study. The definition of the term"study" is not sharp; it has to be emphasized that it is not necessarily identical to a"data set". For example, different data sets may be joined to a multicenter study or may be used to perform a meta-analysis; on the other hand, an existing data set can be reanalysed for a new objective.

When one or more than one statistical test is performed, it is necessary to control type I error(s). If a null hypothesis is rejected, the result is declared"significant". However, one has to distinguish between"scientific" and"probabilistic" significance. For example, if 100 statistical tests are performed, each at the level $\alpha = 5\%$, and 5 "significant" results are obtained, obviously no statistical significance exists;

1990 *Mathematics Subject Classification*. 62H15, 62J15, 62F03.

Key words and phrases. Multiple testing, multiple level, closure test, stochastical dependence among test statistics, logical dependence among hypotheses, model search, hierarchical systems of hypotheses.

We are indebted to Peter Bauer, David Edwards, Irene Schmidtmann and an unknown referee for several suggestions for improving the paper .

in this case a scientific significance could only be asserted, if the objectives corresponding to these"significant" results had been declared as the most important ones in advance. Therefore it is necessary to look for decision rules leading to"real" significances; one solution, which sometimes may be conservative, is the closure principle.

I.2 Multiple test procedures.

The number of tests may be finite or infinite. In the following we will assume, that n test problems, $2 \leq n < \infty$, are given within a parameter space Θ. Each test problem is written as $\Theta = H_i \,\dot{\cup} K_i$, where $H_i \neq \emptyset \neq K_i$, the H_i being the null hypotheses and the K_i the corresponding alternatives. Further, we define the global (null) hypothesis $H_0 = \cap\{H_i : i = 1, \dots, n\}$ and the global alternative $K_0 = \cup\{K_i : i = 1, \dots, n\}$.

A (non-randomized) *multiple test procedure (MTP)* is a family of inferences ϕ_i, where the ϕ_i are functions from the sample space X into the set $\{0, 1\}$. $\phi_i(x) = 0$ indicates retention and $\phi_i(x) = 1$ rejection of H_i, where x is the observed sample result.

I.3 Type I error rates.

For each $\vartheta \in \Theta$ let $F = F_\vartheta$ be the random number of wrong rejections of null hypotheses. One has three common notions of type I error rates (see Miller, 1981; Hochberg/Tamhane, 1987):

Per comparison error rate:	$\text{PCE} = E_\vartheta(F)/n$;
Per-family error rate:	$\text{PFE} = E_\vartheta(F)$.
Familywise error rate:	$\text{FWE} = P_\vartheta(F \neq 0)$
(=*experimentwise error rate,* Tukey, 1953)	

Then, PCE $\leq$ FWE $\leq$ PFE.

An MTP (strongly) controls an error rate at the level α, if it never exceeds α for all $\vartheta \in \Theta$. If an MTP (strongly) controls the FWE, it is often referred to control the *multiple level* (Holm, 1979).

Since the control of merely the PCE can lead to an excessive occurence of type I errors and, if the PFE is controlled, the decisions may become very conservative, it is often recommended to control the FWE. Therefore, in what follows, we mainly consider MTPs controlling the FWE.

Examples:

(1) The"naive" procedure (n tests, each at level α) controls the PCE.

(2) The Bonferroni procedure (n tests, each at level α/n) controls the PFE.

(3) The stepwise procedure of Holm (1979)" $SP(\alpha/n, \alpha/(n-1), \ldots, \alpha/2, \alpha)$" controls the FWE. Here, the symbol" $SP(\alpha_1, \ldots, \alpha_n)$" indicates that the ordered p-values $p_{(i)}$ are compared with α_i. The stepwise rule is that a hypothesis H_j is only rejected, if $p_{(j)} \leq \alpha_j$ *and* $p_{(i)} \leq \alpha_i$ for all $i < j$.

Remark: If the ϕ_i are two-sided tests, it may occur that one concludes correctly that $\phi_i(x) = 1$, but the estimated direction of the difference is wrong. Errors of that kind (sometimes called"type III errors") are very severe, but occur rarely, in general. In the present paper we will not treat the problem how to control type I and type III errors simultaneously.

I.4 Dependences among hypotheses and tests.

There are two kinds of dependences that can be considered within an MTP:

(1) *Stochastical dependences* among test statistics (e.g. a known correlation structure);

(2) *Logical dependences* between hypotheses (e.g. for $H_1 : \mu_1 = \mu_2$, $H_2 : \mu_1 = \mu_3$, $H_3 : \mu_2 = \mu_3$, it can never occur that *exactly* two hypotheses are valid).

Some examples:

(1) The Bonferroni and Holm procedures are general and not specific, since neither of the above mentioned types of dependences are used.

(2) The Tippett (1931) procedure (n tests, each at level $1 - (1 - \alpha)^{1/n}$) assumes that the tests are independent, but needs no information about logical dependences.

(3) The procedure of Ryan (1960) is applicable for all pairwise comparisons of parameters of any kind. On the other hand, the tests and their stochastical dependences are arbitrary.

(4) The classical MTPs in the ANOVA model (see Tukey, 1953; Scheffé, 1953; Dunnett, 1955) use information about both kinds of dependences.

II. THE CLOSURE PRINCIPLE

II.1 Stepwise procedures in the ANOVA model.

In the standard one-way ANOVA set, we suppose that there are $k \geq 3$ treatments and $n_i \geq 2$ independent replications for each treatment. We consider the model

$$X_{i\ell} = \mu_i + \epsilon_{i\ell}$$

with $\epsilon_{i\ell} \sim N(0, \sigma^2)$ and independently distributed for $i = 1, \ldots, k$ and $\ell = 1, \ldots, n_i$. Our interest is to test all $\binom{k}{2}$ pairwise comparison hypotheses

$$H_{ij} : \mu_i = \mu_j, \qquad 1 \leq i < j \leq k,$$

and we require that the MTP controls the FWE.

There are some well-known tests for the global hypothesis $H_0 = \cap H_{ij}$:

(1) the classical F test which is based on a (weighted) variance of the sample means $\bar{x}_i$;
(2) the studentized range test, based on $R = \max\{| \bar{x}_i - \bar{x}_j |: 1 \leq i < j \leq k\}$;
(3) the global Bonferroni test which compares the smallest p-value of the $\binom{k}{2}$ t tests at level $\alpha/\binom{k}{2}$.

From these global tests, the following multiple test procedures are derived:

(1) the procedure of Scheffé (1953) which can be applied not only to pairwise comparisons, but also to linear contrast hypotheses;
(2) the"honestly significant difference" (HSD) test of Tukey (1953);
(3) the procedure of Dunn (1961).

All procedures are *coherent*, which means that an H_{ij} is only rejected, if H_0 has been rejected. Moreover, the procedures 2. and 3. are *consonant*, i. e. when H_0 is rejected, one can reject at least one H_{ij}. All procedures control the FWE, but are not stepwise; it can be expected that there are more powerful stepwise procedures controlling the FWE.

The first proposal of a stepwise procedure was Fisher's (1935) LSD test: At first, one performs an F test at global level α; if H_0 is retained, the procedure has to be stopped; otherwise one continues with t tests (at level α) for all pairwise comparisons. For $k = 3$, the LSD test controls the FWE, but the FWE α can be exceeded for $k \geq 4$ (e.g. for $\mu_1 = \mu_2 = \mu_3 \neq \mu_4$). To remedy this, Newman (1939) and Keuls (1952) proposed a strategy with more steps, which considers all hypotheses of homogeneity. But even for this procedure the FWE may exceed α (e.g. for $\mu_1 = \mu_2 \neq \mu_3 = \mu_4$). Hence, within the stepwise procedure, all ***partition*** hypotheses have to be considered, too (see Cox/Spjøtvoll, 1982); and it has to be ensured that the whole procedure is ***coherent***, i.e. a hypothesis can only be rejected if all hypotheses implying it are also rejected.

A generalization of this principle will be formulated in the next section.

II.2 The specific closure method.

We assume that the hypotheses H_i and tests ϕ_i, $i = 1, \ldots, n$, satisfy the following conditions:

(1) The system $\mathcal{H} = \{H_1, \ldots, H_n\}$ is *closed under intersection*, i.e. $H_i \cap H_j \in \mathcal{H}$ for all i, j.
(2) The tests ϕ_i are *tests at level* α for H_i, i.e. for $\vartheta \in H_i$ one has $P_\vartheta(\phi_i = 1) \leq \alpha$.
(3) The MTP $(\phi_1, \ldots, \phi_n)$ is *coherent*, i.e. for $H_i \subseteq H_j$ and $\phi_i = 0$ it follows $\phi_j = 0$.

Under these conditions, the MTP $(\phi_1, \ldots, \phi_n)$ controls the FWE at the level α (Marcus et al., 1976; Sonnemann, 1982).

In many practical situations, the given system $\mathcal{H}$ is not a priori closed under intersection. In this case, one can form $\mathcal{H}^c$ = the closure of $\mathcal{H}$, which is the system of all intersections of hypotheses from $\mathcal{H}$. Then, one needs level α tests ϕ_H for all hypotheses $H \in \mathcal{H}^c$. This procedure may lead to a substantial inflation of $\mathcal{H}$, since $\mathcal{H}^c$ may consist of up to $(2^n - 1)$ hypotheses.

A possible test for an intersection hypothesis $H = H_{i_1} \cap \cdots \cap H_{i_m}$ is the"global" *Bonferroni test*, i.e. $\phi_H = 1$, iff at least one of the $H_{i_j}, j = 1, \ldots, m$, can be rejected at the level α/m.

Moreover, the MTP $(\phi_H : H \in \mathcal{H}^c)$ need not be coherent. In this case, one can *enforce coherence* by introducing a new MTP $(\psi_H : H \in \mathcal{H}^c)$, where $\psi_H = \inf\{\phi_{H'} : H' \subseteq H, H' \in \mathcal{H}^c\}$. As a result, $\mathcal{H}^c$ and $(\psi_H : H \in \mathcal{H}^c)$ satisfy the above conditions, and the FWE is controlled at the level α.

II.3 A modified closure method.

The closure method, as described in II.2, is ***specific*** for the given multiple test problem and can be expected to be powerful. However, the inflation of $\mathcal{H}$ as well the enforcement of coherence may lead to substantial algorithmic problems, even if a computer is used. Therefore, it is useful to investigate *general* concepts which can be applied even for large systems of hypotheses. In the following we consider an implication-like structure (the same for all multiple test problems), which does not emphasize the inclusion of hypotheses, but rather of index sets $I \subseteq \{1, \ldots, n\}$:

Given is the system $\mathcal{H} = \{H_1, \ldots, H_n\}$, not necessarily closed under intersection. For all $I \subseteq \{1, \ldots, n\}, I \neq \emptyset$, the intersection hypotheses are defined as $H_I = \cap\{H_i : i \in I\}, H_\emptyset = \Theta$. Assumed that the MTP $(\phi_I : I \subseteq \{1, \ldots, n\}, I \neq \emptyset)$ is *quasi-coherent* (i.e. if $I \supseteq J$ and $\phi_I = 0$, it follows that $\phi_J = 0$), then it controls the FWE at the level α (Hommel,1986).

Similarly as in II.2, one can enforce quasi-coherence by introducing an MTP $(\psi_I : I \subseteq \{1, \ldots, n\}, I \neq \emptyset)$, where $\psi_I = \inf\{\phi_J : J \supseteq I\}$.

II.4 General global tests and multiple tests.

In order to apply the method described in II.3, one needs a *global* test ϕ_I for each H_I. We assume that there are given *elementary* tests for each H_i, the results of which can be represented as p-values $P_i, i = 1, \ldots, n$. The ordered p-values $P_i, i \in I$, are denoted by $P_{1:I} \leq P_{2:I} \leq \ldots \leq P_{|I|:I}$. The following tests are global tests of H_I:

(1) The global Bonferroni test: Reject H_I iff $P_{1:I} \leq \alpha / \mid I \mid$.
(2) Rüger's (1978) test: Let $k, 1 \leq k \leq \mid I \mid$, be fixed in advance. Reject H_I iff $P_{k:I} \leq k \cdot \alpha / \mid I \mid$.
(3) The combination of Rüger's tests (Hommel, 1983): Reject H_I iff there exists a $k, 1 \leq k \leq \mid I \mid$, such that $P_{k:I} \leq k \cdot \alpha / (\mid I \mid \cdot C_{|I|})$, where $C_m = 1 + 1/2 + \cdots + 1/m$ for $m \geq 1$.
(4) The Simes (1986) test: Reject H_I iff there exists a $k, 1 \leq k \leq \mid I \mid$, such that $P_{k:I} \leq k \cdot \alpha / \mid I \mid$.

It has been shown that the tests 1.-3. are tests at level α. For Simes' test, a mathematical proof has been given only in case of independent tests. However, simulation studies of Simes (1986) and of Hommel (1988) suggest that the level α is still kept, if the p-values are based on common multivariate distributions, e.g. multivariate normal or multivariate χ^2 distributions.

By application of the modified closure method using the above mentioned global tests ϕ_I, one obtains the following MTPs:

(1) If the ϕ_I's are global Bonferroni tests, one obtains Holm's (1979) procedure.
(2) If all ϕ_I's are Rüger tests or

(3) combinations of Rüger tests, some arising procedures have been described by flow charts (Hommel, 1986).

(4) If the ϕ_I's are Simes tests, the MTP derived can also be described by a flow chart. However, for the elementary hypotheses H_i the decision procedure is much simpler (Hommel, 1988). One has to calculate $j = \max\{i \in \{1, \ldots, n\} : P_{(n-i+k)} > k \cdot \alpha/i$ for $k = 1, \ldots, i\}$ and can reject all H_i with $P_i \leq \alpha/j$. (If the maximum does not exist, one can even reject *all* H_i, $i = 1, \ldots, n$).

The FWE α is controlled for the above MTPs 1., 2. and 3. For the multiple Simes test 4. the control of the FWE is ensured only if the tests ϕ_I are actually tests at level α.

III. IMPROVED MTPs USING DEPENDENCIES

III.1 Use of stochastical dependencies.

If it is known that the p-values are independent, one can apply the following global tests ϕ_I at level α for each H_I:

(1) Fisher's test (1925):
Reject H_I iff $-2 \cdot \ln \prod_{i \in I} P_i \geq \chi^2_{2 \cdot |I|; \alpha}$.

(2) The Tippett test (1931):
Reject H_I iff $P_{1:I} \leq 1 - (1 - \alpha)^{1/|I|}$.

(3) The Simes test (see II.4).

(4) Maurer's test (see Maurer/Mellein, 1988):
Reject H_I iff $P_{1:I} \leq \alpha \cdot (1 - P_{|I|:I})/(1 - \alpha)$.

If the modified closure method is applied, one obtains the following simple MTPs:

(2) The Tippett-Holm procedure:
$SP(1 - (1 - \alpha)^{1/n}, 1 - (1 - \alpha)^{1/(n-1)}, \ldots, 1 - (1 - \alpha)^{1/2}, \alpha)$.

(4) The multiple Maurer test:
Reject H_i iff $P_i \leq \alpha \cdot (1 - P_{(n)})/(1 - \alpha)$.

The multiple Simes test has been described in II.4, whereas an MTP using Fisher tests seems to become much more complicated.

If the p-values are non-independent, one can use information about the dependence structure. In case that the distribution of the test statistics for each H_I is known, one can find specific level α tests. For instance, in the ANOVA model one can use F tests or studentized range tests. Non-specific solutions are Šidák-style adjustments (the same bounds as for Tippett's global test; see Šidák, 1967) or, again, application of the Simes test. Moreover, one can apply in some situations

Bonferroni inequalities of higher order (see Hunter, 1976). However, for all these techniques the problem remains that the arising MTPs should not become too complicated.

III.2 Use of logical dependencies.

In this section, we summarize the results of Bergmann/Hommel (1988).

An index set I is called *exhaustive*, if H_I is non-empty and for all $J \subseteq \{1, \ldots, n\}$ with $H_I = H_J$ it follows that $I \supseteq$ J. $\mathcal{P}_{EI}$ is the set of all exhaustive index sets. EL is the set of all exhaustive"layers", i.e. the set of all $i, 1 \leq i \leq n$, for which at least one $I \in \mathcal{P}_{EI}, \mid I \mid = i$, exists. Further, let $t_i = \max\{j \in EL; j \leq n - i + 1\}, 1 \leq i \leq n$.

Example.

Consider all pairwise comparisons of 4 parameters μ_1, μ_2, μ_3, μ_4 with the 6 hypotheses

$H_1 : \mu_1 = \mu_2$, $\quad H_2 : \mu_1 = \mu_3$, $\quad H_3 : \mu_1 = \mu_4$,
$H_4 : \mu_2 = \mu_3$, $\quad H_5 : \mu_2 = \mu_4$, $\quad H_6 : \mu_3 = \mu_4$.

Then, the following index sets are exhaustive:

$\{1, \ldots, 6\}$, $\{1,2,4\}$, $\{1,3,5\}$, $\{2,3,6\}$, $\{4,5,6\}$, $\{1,6\}$, $\{2,5\}$, $\{3,4\}$, $\{1\}, \ldots, \{6\}$.

Since there are no exhaustive sets with four or five elements, one obtains $EL = \{1,2,3,6\}$, and therefore $t_1 = 6$, $t_2 = t_3 = t_4 = 3$, $t_5 = 2$, $t_6 = 1$.

In the paper of Bergmann/Hommel four different improvements of Holm's procedure (which is called"Procedure 0" by the authors) are given, which are based on global Bonferroni tests and are, in ascending sequence, more and more powerful.

"Procedure 1" is Shaffer's (1986) "modified sequentially rejective Bonferroni" (MSRB) procedure, which can be written as $SP(\alpha/t_1, \ldots, \alpha/t_n)$.

"Procedure 2" of Bergmann/Hommel is only important for unequal weights. If the weights for the tests are identical, as in the present paper, it coincides with Shaffer's procedure.

"Procedure 3": A further improvement of Shaffer's MSRB can be attained, if at each step of the procedure the hypotheses not yet rejected are examined more individually. One introduces a "rejection set" R which is the set of all indices i, for which the corresponding H_i can be rejected. At the beginning of the stepwise procedure, R is empty. In a typical step, it is checked whether there exists a $j \notin R$ with $P_j \leq \alpha/m$, where $m = \max\{| I |: I \in \mathcal{P}_{EI}, I \cap R = \emptyset, j \in I\}$. In this case, one forms a new rejection set $R := R \cup \{j\}$, and the typical step is repeated; otherwise the procedure stops with a final set R.

"Procedure 4" is still more powerful than Procedure 3. It has been presented by Bergmann/Hommel not as a fully elaborated algorithm. The principle of the procedure is to determine an "acceptance set"

$$A = \cup\{I : I \in \mathcal{P}_{EI}, P_{1:I} > \alpha / \mid I \mid\}.$$

Then all H_j with $j \notin A$ can be rejected. The decisions of this procedure and the specific closure test using Bonferroni global tests coincide for all $H_j, 1 \leq j \leq n$; but situations may occur, where an H_I with $| I |> 1$ is not rejected by Procedure 4, but by the latter one.

For the above procedures, only"global" Bonferroni tests ϕ_I were used. However, the Bonferroni inequality may be very rough (in particular for model search problems), and one should find more powerful tests ϕ_I so that the above mentioned testing strategies are still applicable. This is often possible by Šidák-style adjustments or, in the ANOVA situation, by studentized range tests.

It is also possible to apply Simes global tests instead of Bonferroni tests. In analogy to Shaffer's procedure, one obtains an improvement of the multiple Simes procedure described in II.4 if j is determined by

$$j = \max\{i \in EL : P_{(n-i+k)} > k \cdot \alpha/i \text{ for } k = 1, \ldots, i\}$$

and all H_i with $p_i \leq \alpha/j$ are rejected. For Procedures 3 and 4 one also gets feasible improvements, if the application of Simes tests is allowed; however, for finding optimal strategies, further investigation is necessary.

For sake of simplicity, we will describe in the rest of the paper only"Bonferroni-type" procedures and algorithms which are applicable for arbitrary kinds of stochastical dependencies, but often can be improved in special situations.

III.3 A rapid algorithm for Procedure 3 and 4.

To perform Procedure 3 or 4 one needs the knowledge of all exhaustive index sets. This set can simply be determined, if all subsets of $\{1, \ldots, n\}$ are checked whether they are exhaustive or not. However, for larger n this procedure is very time-consuming. In a system of hypotheses with a greater number of logical dependences the expense of checkings can be substantially reduced using an algorithm proposed by Bernhard (1992). A flow chart of the algorithm is given in Figure 1. For this purpose the following notations are used:

$$\mathcal{P}_n = \text{the power set of } \{1, \ldots, n\}$$

$$\mathcal{P}_{EI} = \text{the set of all exhaustive index sets.}$$

For $I \in \mathcal{P}_n$, we denote

$$SE(I) = \text{the smallest exhaustive index set containing } I$$

$$i^* := i^*(I) := \max(I) \qquad \text{if} \quad |I| \geq 1$$

$$i^{**} := i^{**}(I) := \max(I \backslash \{i^*\}) \qquad \text{if} \quad |I| \geq 2$$

and

$$I^R := I^R(I) := \begin{cases} (I \backslash \{i^*, i^{**}\}) \cup \{i^{**} + 1\} & \text{if} \quad |I| \geq 2 \\ \emptyset & \text{if} \quad |I| = 1 \end{cases}$$

Furthermore the lexicographical order on $\mathcal{P}_n$ is defined by the following recursive formula

$$I_1 := \{1\}$$

$$I_{k+1} := \begin{cases} I_k \cup \{i^*(I_k) + 1\} & \text{if} \quad i^*(I_k) \leq n \\ I^R(I_k) & \text{if} \quad i^*(I_k) = n \end{cases}$$

for $k = 1, \ldots, (2^n - 1)$.

The algorithm is based on the idea to determine $SE(I)$ for a given I, instead of verifying, whether an index set I is exhaustive. Thus, in each step regarding an $I \in \mathcal{P}_n$ one always gets an exhaustive index set. Since the algorithm is founded upon the lexicographical order the criteria $C1$ und $C2$ ensure that each $I \in \mathcal{P}_{EI}$ is obtained exactly once. Criterion $C1$ checks, whether the calculation of $SE(I)$ is necessary. Criterion $C2$ decides, whether the calculated index set $SE(I)$ is really a new exhaustive index set. Since most of the computing time is required to calculate $SE(I)$ the improved algorithm reduces the expense of computing time.

Example.

Consider all pairwise comparisons of p parameters, then $n = \binom{p}{2}$ elementary hypotheses exist. Using the simple strategy one has to check $2^n - 1$ index sets. For $p = 8$ (6) parameters one obtains $n = 28$ (15) hypotheses, $|\mathcal{P}_n| = 2^{28} - 1 = 2.7 \cdot 10^8$ ($2^{15} - 1 = 32 \cdot 10^3$). The number of determinations of $SE(I)$ is about $3 \cdot 10^4$ (884), whereas $|\mathcal{P}_{EI}| = 4.139$ (202).

To perform Procedure 3 or 4 using the improved algorithm one needs, in addition, an algorithm to calculate $SE(I)$ for a given I. If the hypotheses can be represented as linear combinations of some parameters

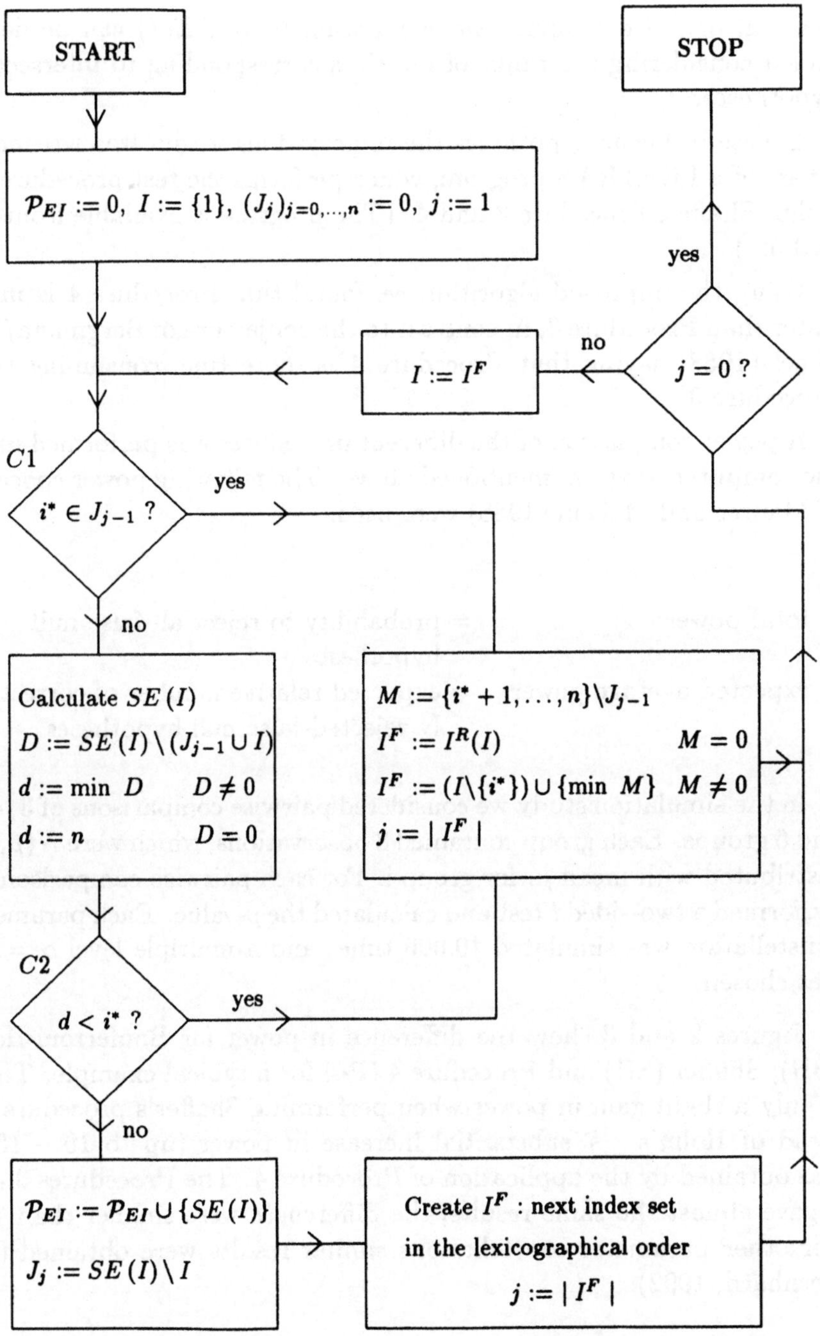

Figure 1: Flow chart of an improved strategy to determine $\mathcal{P}_{EI}$.

(e.g. all pairwise comparisons of p parameters), $SE(I)$ can be determined considering the ranks of matrices corresponding to intersection hypotheses.

In case of linear hypotheses the improved algorithm was written as a part of a FORTRAN program, which performs the test procedures of Holm, Shaffer, Procedure 3 and 4. (The program is available from the authors).

Using the improved algorithm we found that Procedure 4 is much faster than Procedure 3, in contrast to the conjecture of Bergmann/Hommel (1988) saying that Procedure 4 is more time consuming than Procedure 3.

A power comparison of the different procedures was performed using the computer program mentioned above. The following power concepts of Maurer and Mellein (1988) were used:

total power	= probability to reject all false null hypotheses
expected average power	= expected relative number of correctly rejected false null hypotheses.

In the simulation study we considered pairwise comparisons of 3, 4, 5 and 6 groups. Each group contained 8 observations, which were $N(\mu_i, 1)$ distributed with mean μ_i for group i. For each pairwise comparison we performed a two-sided t test and calculated the p-value. Each parameter constellation was simulated 10.000 times and a multiple level $\alpha = 5\%$ was chosen.

Figures 2 and 3 show the difference in power for Bonferroni-Holm (BH), Shaffer (SH) and Procedure 4 (P4) for a typical example. There is only a slight gain in power when performing Shaffer's procedure instead of Holm's. A substantial increase in power (up to $10 - 15\%$) was obtained by the application of Procedure 4. The Procedures 3 and 4 gave almost the same results, the differences were smaller than 1%. For other parameter constellations similar results were obtained (see Bernhard, 1992).

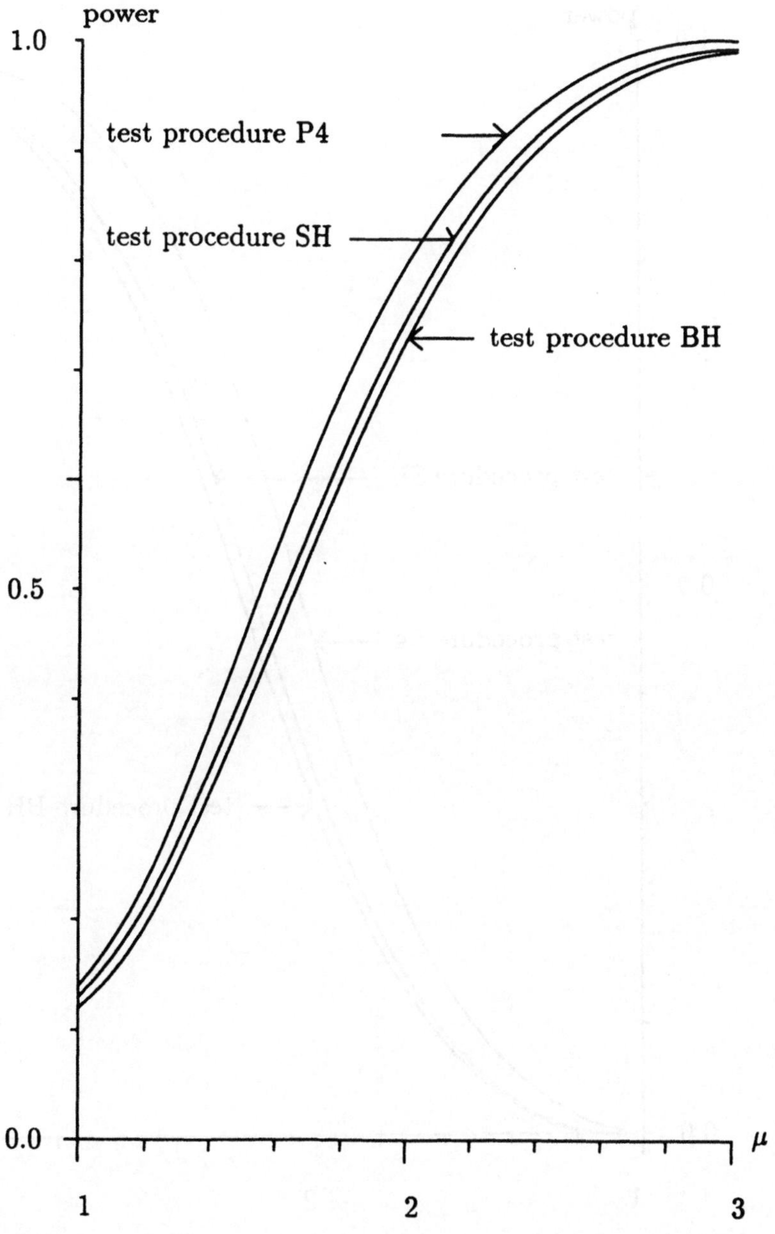

Figure 2. Expected average power of Holm(BH), Shaffer(SH) and Procedure 4 (P4) for the pairwise comparison of 6 groups with means $(0, 0, 0, 0, \mu, \mu)$.

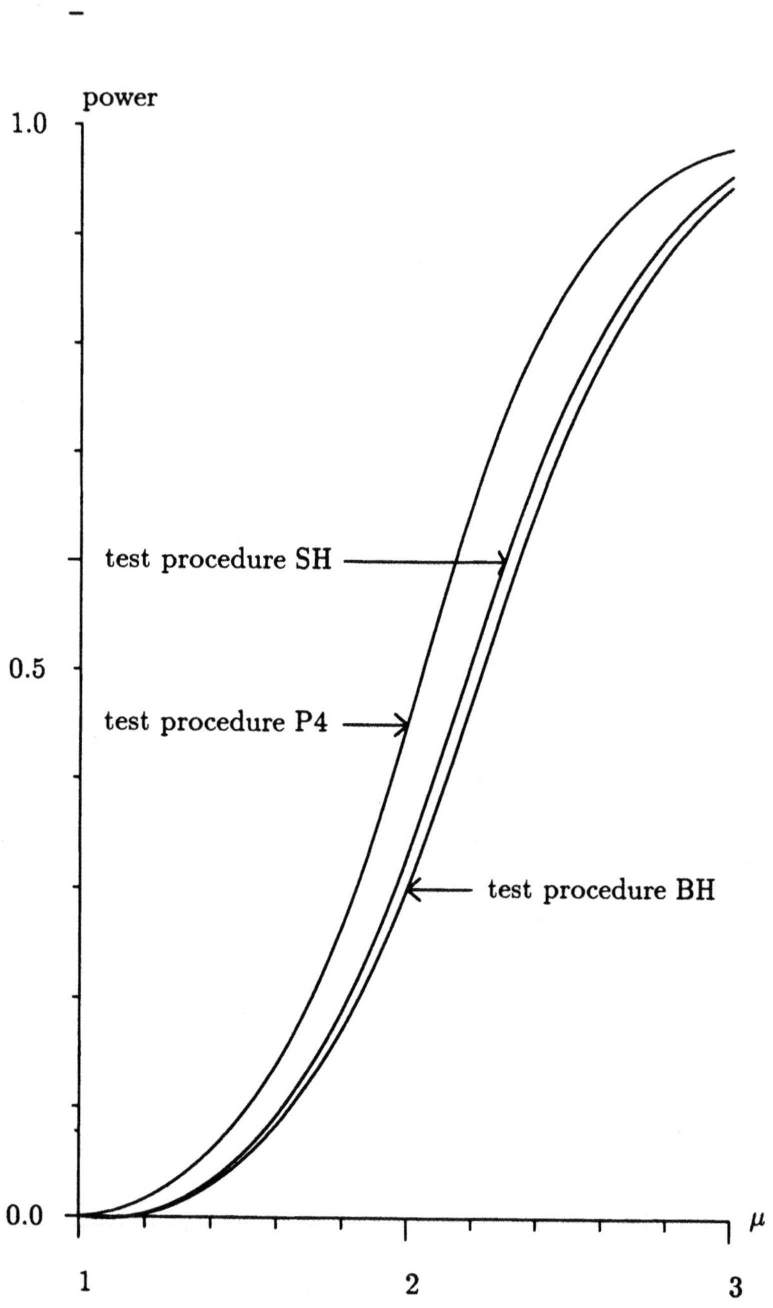

Figure 3. Total power of Holm(BH), Shaffer(SH) and Procedure 4 (P4) for the pairwise comparison of 6 groups with means $(0, 0, 0, 0, \mu, \mu)$.

IV. MULTIPLE TESTING IN MODEL SEARCH

IV.1 General problems.

We assume that the parameter space Θ consists of the set of all possible models. One can often choose Θ = all m-dimensional real vectors, with $m \geq 1$.

Example 1. (Gabbert et al., 1992)

In 442 patients with gastric cancer factors were searched, which might be of prognostic relevance for the time of survival. The statistical analysis was performed by means of Cox's regression (cf. Kalbfleisch/Prentice, 1980, pp.70). The following variables were considered:

T: Tumor stage (T1-T4)

N: Lymph node involvement (yes/no)

G: Grade of tumor differentiation (G1-G4)

TCD: Degree of tumor cell dissociation (0-3).

The set of all models can be described by the hazard function

$$h(t) = h_0(t) \cdot \exp\{\beta_T \cdot T + \beta_N \cdot N + \beta_G \cdot G + \beta_{TCD} \cdot TCD\},$$

with the parameter space

$$\Theta = \{(\beta_T, \beta_N, \beta_G, \beta_{TCD}) : -\infty < \beta_i < +\infty\} \text{ ("T+N+G+TCD").}$$

The question was whether, in addition to T, the node state N is informative for prognosis. A further question was whether one or both of the variables G and TCD are useful for a better prognosis than using T and N alone. Therefore the model hypotheses are

H_1: $\beta_N = \beta_G = \beta_{TCD} = 0$ ("T");

H_2: $\beta_G = \beta_{TCD} = 0$ ("T+N");

H_3: $\beta_{TCD} = 0$ ("T+N+G");

H_4: $\beta_G = 0$ ("T+N+TCD").

It is typical for model search problems that the structure of the model hypotheses is completely or partially hierarchical, and that the model hypotheses consist of linear functions of the coordinates of Θ and can therefore be represented as matrices. The aim is to choose valid or at least"almost valid" models, but the models should be parsimonious, i.e. they should be as simple as possible (however, no unique solution need exist for this purpose).

Some procedures for model choice which are not based on multiple hypotheses testing, but have the objective to select parsimonious models, are Bayes procedures (Box/Hill, 1967), the C_p criterion of Mallows (1973) or the AIC criterion of Akaike (1974). An important property is the"consistency" of the respective procedures, i.e. the probability of selecting the most parsimonious correct model tends asymptotically to 1. A solution of this problem using methods of multiple hypotheses testing explicitly has been given, under rather weak conditions, by Bauer/P"otscher/Hackl(1988); the concept is that α tends asymptotically to zero with suitable convergence speed.

In our context, the model hypotheses are considered as null hypotheses; then the control of any error rate α causes primarily that the accepted models are not too complicated, which could be meaningful if costs for future measurements have to be saved. However, one might prefer to control rather type II than type I errors (i.e. one prefers a true model even if it is too complicated). In this case, a problem arises which is typical, if null and alternative hypotheses are exchanged, namely that the model hypotheses have to be formulated as interval hypotheses.

When type I errors have to be controlled, it is possible that α is not chosen as a fixed level, but that rather an *adjusted p-value* is computed, i.e. the smallest significance level for which the corresponding model hypothesis can be rejected (cf. Westfall/Young, 1989). It can be considered as a descriptive measure for the"plausibility" of a model hypothesis.

Further, it should be noted that control of the FWE is not as desirable as in many other situations, but if only the"naive" procedure is applied (which controls the PCE) the decisions may become incoherent.

IV.2 Decompositions of test statistics for completely hierarchical model sequences.

There are two general techniques which can be employed for the choice of suitable models: (i) likelihood ratio tests, and (ii) decomposition of the sum of squares. In particular, for both techniques a good interpretation is possible for each completely hierarchical sequence of models, since one obtains a decomposition of the test statistics for the whole sequence into independent components.

Example 1. (continued)

A complete hierarchical sequence of models is T $\rightarrow$ T+N $\rightarrow$ T+N+G $\rightarrow$ T+N+G+TCD; under H_1, the corresponding three χ^2 tests are asymptotically independent, and their sum 4.57+1.86+5.35=11.78

is the test statistic for the test of H_1 within the full model space Θ (under H_1, this sum is asymptotically χ^2-distributed with 3 d.f.). Moreover, the sum of the χ^2"arrow test" statistics is the same for all possible ways (for example, 1.86+5.35=6.94+0.27=7.21 for the way T+N $\rightarrow$ T+N+G+TCD).

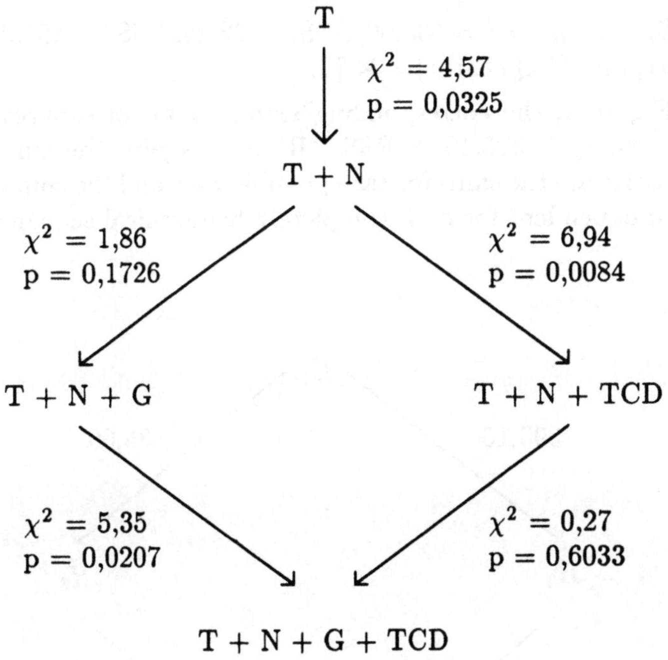

Figure 4.Results of the"arrow tests" for a restricted patient collective of Example 1. Each χ^2 test has 1 d.f. and can be interpreted as a test of the model at the beginning of the arrow within the (restricted) model space at the end of the arrow.

Example 2. (Hartung et al., 1987, p. 598 ff.)

For 10 work pieces the dependence of working time (y) on diameter (x_1), length (x_2), and additionally on the product $x_1 \cdot x_2$ was investigated.

The full model is written as

$$Y_i = \beta_0 + \beta_1 x_{1i} + \beta_2 x_{2i} + \beta_3 x_{1i} x_{2i} + \epsilon_i,$$

with $\epsilon_i \sim N(0, \sigma^2)$ and independently distributed for $i = 1, \ldots, 10$.

For the interesting model hypotheses

$$H_1 : \beta_1 = \beta_2 = \beta_3 = 0$$
$$H_2 : \beta_2 = \beta_3 = 0$$
$$H_3 : \beta_1 = \beta_3 = 0$$
$$H_4 : \beta_3 = 0$$

one obtains the following residual sums of squares:

$\mathrm{RSS}_1 = \Sigma(y_i - \overline{y})^2 = 468.00$, $\mathrm{RSS}_2 = 72.84$, $\mathrm{RSS}_3 = 431.34$, $\mathrm{RSS}_4 = 34.97$, and SSQ (Error)= 34.72.

In Figure 5, the corresponding"arrow" sums of squares are given (for example, 395.16 = RSS_1- RSS_2). Again, the sums of these statistics are the same for each possible way, and the sums of squares are independent for each completely hierarchical sequence.

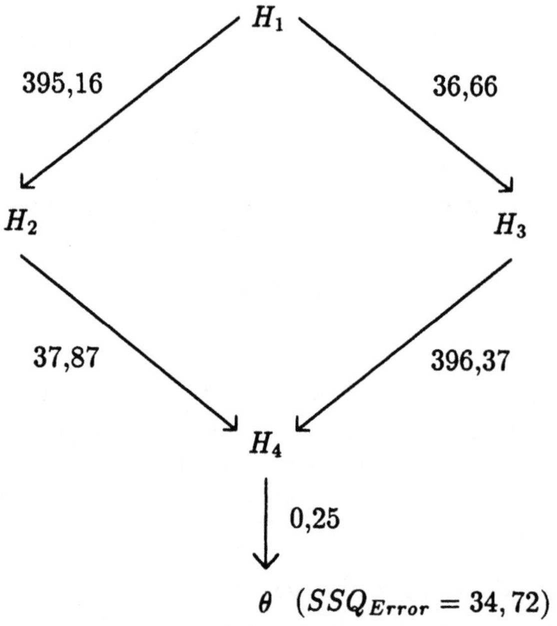

Figure 5.The "arrow" sums of squares for Example 2.

i	1	2	3	4	5	6	7	8	9	10
x_{1i}	4	6	5	3	5	3	3	6	4	5
x_{2i}	31	3	7	29	27	18	28	17	8	12
y_i	24	28	25	13	27	14	15	32	14	28

Data for the example 2.

IV.3 Completely hierarchical systems of hypotheses.

Sometimes the structure of a given system of model hypotheses is completely hierarchical, e.g. if the degree of a polynomial has to be determined. Let

$$Y_i = \beta_0 + \beta_1 x_i + \beta_2 x_i^2 + \cdots + \beta_n x_i^n + \epsilon_i$$

with $\epsilon_i \sim N(0, \sigma^2)$ and independently distributed for $i = 1, \ldots, N$; then the hypotheses of interest might be

$$H_k : \beta_k = \beta_{k+1} = \cdots = \beta_n = 0 \quad \text{for } k = 1, \ldots, n.$$

If we denote S_{k-1}^2 the residual sum of squares under model H_k for $k = 1, \ldots, n$ and $S_n^2 =$ SSQ (Error) for the full model, then the usual F tests of H_k are performed using the test statistics

$$F_k = \frac{(S_{k-1}^2 - S_n^2)/(n-k+1)}{S_n^2/(N-n-1)}$$

with $(n-k+1; N-n-1)$ d.f.

Now, one can simply perform the closure test using the test statistics F_k. Another way is to apply Procedure 4. Since $SE(\{k\}) = \{k, k+1, \ldots, n\}$, the tests for H_k are global Bonferroni tests using $F_k, F_{k+1}, \ldots, F_n$. The resulting multiple test procedure is a Bonferroni-Holm procedure with stepwise bounds α/m, where m is the number of hypotheses not rejected up to this step (Bauer/Hackl, 1987). However, since the F_i may be highly dependent, the application of this procedure does not seem to be powerful. But instead of F_k (where H_k is tested against $\Theta \backslash H_k$) one can use, according to a recommendation of Bauer/Hackl, the test statistics

$$F_k' = \frac{S_{k-1}^2 - S_k^2}{S_k^2/(N-k-1)}$$

testing H_k against $H_{k+1} \backslash H_k$, which are, under H_k, F-distributed with $(1; N-k-1)$ d.f. For the F'_k, the numerators $(S^2_{k-1} - S^2_k)$ are independent, and in the denominators one can save degrees of freedom (for $k < n$), in particular when N is only slightly larger than n. It should be noted that it is insufficient to perform the closure test using the test statistics F'_k because, for $k < n$, the tests based on F'_k are not consistent for some $\vartheta \in \Theta \setminus H_k$.

The general idea, that in a completely nested system a hypothesis H_k is tested not within the full parameter space Θ, but within the restricted space H_{k+1}, has also been treated by Alt (1988).

A simulation study demonstrated that indeed Procedure 4 using F_k tests cannot be recommended. On the other hand, if F'_k tests are used, Procedure 4 is, in general, more powerful than the closure test using F_k tests, especially when S^2_n has only few degrees of freedom.

IV.4 Partially hierarchical systems of hypotheses.

When the system of hypotheses is not completely, but partially hierarchical, as in Examples 1 and 2, one can use similar ideas as in IV.3. Of course, it is possible to apply the closure test with tests within the full parameter space (for the example of loglinear models see Schiller, 1988); in the following, the term"closure test" is used for MTPs of this type. Moreover, one can use several"Bonferroni-type" procedures, for which the procedures described in III.2 and, for Procedure 3 and 4, the algorithms described in III.3 can be applied. For these procedures, different sets of elementary hypotheses can be considered; furthermore, the structure of the elementary tests ϕ_i has to be specified (in particular, whether the ϕ_i are tests within the full or within a restricted parameter space).

We call a test"saturated" if it is a test within the full space Θ (e.g. the tests based on F_k in IV.3). A test is called an"arrow test" if the parameter space is one of the smallest models which are more complex than the null hypothesis (as the tests based on F'_k in IV.3).

We have investigated the following three Bonferroni-type strategies:

S1: The elementary hypotheses are the model hypotheses, the elementary tests are saturated tests as used by the closure test.

S2: A"parameter-oriented" strategy: The elementary hypotheses are only those model hypotheses which are the most complex ones setting a specific parameter to zero. The elementary tests are arrow tests leading to the closest model(s) where this parameter is allowed to be different from zero. In many practical situations, this definition of an elementary test is unique; if it is not, one can use similar ideas as described for the following strategy S3.

S3: A complete"arrow-oriented" strategy: The elementary hypotheses are all hypotheses at the beginning of an arrow (indicating the direction of the implication) in the corresponding diagram. If more than one arrow proceeds from a model, the respective hypotheses are considered as distinct. The elementary tests are the analogous arrow tests.

If the closure test and the strategies S1, S2, S3 (using Procedure 4, which is the most powerful one of the Bonferroni-type procedures) are applied to the data of Example 1, one obtains the following results (using a multiple level of $\alpha = 0.05$):

Closure test: At first, H_1 is tested: $\chi^2 = 11.78$ with 3 d.f., $p_1 = .0082 < \alpha$. Therefore, one can reject H_1 and continue with testing $H_2 : \chi^2 = 7.21$ with 2 d.f., $p_2 = .0272 < \alpha$, and H_2 is rejected. Hence, H_3 and H_4 can be tested each at the level α: For $H_3 : \chi^2 = 5.35$ with 1 d.f., $p_3 = .0207 < \alpha$, whereas for H_4, $p_4 = .6033 > \alpha$.
By the closure test the hypotheses H_1, H_2, H_3 are rejected, H_4 is retained, and"T+N+TCD" is a plausible model.

S1: From Procedure 4, using the same 4 p-values p_1, p_2, p_3, p_4 as before, one obtains in the first step: $\min\{p_1, p_2, p_3, p_4\} = p_1 = .0082 < \alpha/4$ and therefore H_1 is rejected. But since $| SE(\{2\}) | =| \{2,3,4\} |= 3$ and $\min\{p_2, p_3, p_4\} > \alpha/3$, one has to retain H_2, H_3, H_4, and"T+N" would be a plausible model.

S2: The three parameters of interest are β_N, β_{TCD} and β_G, and one has to consider three elementary hypotheses, too. The corresponding three arrow tests are given
- for β_N: use $H_1 \rightarrow H_2$, with p'_1 =.0325
- for β_{TCD}: use $H_3 \rightarrow \Theta$, with p'_2 =.0207
- for β_G: use $H_4 \rightarrow \Theta$, with p'_3 =.6033

Since $\min\{p'_1, p'_2, p'_3\} > \alpha/3$, no hypothesis can be rejected, and this procedure would declare"T" as a plausible model.

S3: We consider all five arrows in Figure 4, each corresponding to a separate elementary hypothesis. Therefore the following five arrow tests are employed:

$$H_1 \rightarrow H_2 : p''_1 = .0325;$$

$$H_2 \rightarrow H_3 : p''_2 = .1726;$$

$H_2 \rightarrow H_4 : p_3'' = .0084;$

$H_3 \rightarrow \Theta : p_4'' = .0207;$

$H_4 \rightarrow \Theta : p_5'' = .6033.$

Since $p_3'' < \alpha/5$, one has $\min\{p_i'' : i \in I\} < \alpha / \mid I \mid\}$ for all exhaustive index sets I with $3 \in I$, and 3 is not an element of the acceptance set A. Hence one can reject the model hypothesis corresponding to p_3'', i.e."T+N" is rejected, and therefore the hypothesis implying"T+N", i.e."T", is automatically rejected. What follows is that the elementary hypotheses corresponding to p_1'', p_2'', p_3'' are rejected, and only p_4'' and p_5'' remain. Since $SE(\{4,5\}) = \{2,3,4,5\}$ (the indices referring to those of p_i''), and the corresponding hypothesis has been rejected, one only has to consider $SE(\{4\}) = \{4\}$ ($p_4'' < \alpha$, the model"T+N+G" is rejected) and $SE(\{5\}) = \{5\}$. Since $p_5'' > \alpha$, one obtains"T+N+TCD" as a plausible model, like in the closure test.

Remark: This solution can be found more formally, if it is verified that there exist only 4 exhaustive index sets, namely $\{1,2,3,4,5\}$, $\{2,3,4,5\}$, $\{4\}$ and $\{5\}$.

As mentioned above, by this strategy the same hypothesis can be tested by means of different tests; in our example, H_2 ("T+N") is tested within the restricted model spaces H_3 (leading to p_2'') or H_4 (leading to p_3''). Of course, rejection by at least one of these arrow tests is sufficient for the rejection of the model hypothesis.

Because Procedures 3 and 4 can be employed for each of the strategies S1, S2, S3, the computer program using the algorithm described in III.3 can be used. Since the test statistics for strategy S1 (as in IV.3) can be highly correlated, it cannot be expected that this strategy is powerful. However, in contrast to completely nested systems, it is not possible, also for strategies S2 and S3, to obtain a complete decomposition into independent test statistics. First simulations showed, in fact, that the closure test has the highest power, in general, whereas strategies S1 and S2 often have small power. Strategy S2 is attractive, first of all, because of its simplicity. Strategy S3 is recommended when there are only few ramifications within the system of hypotheses, i. e. it is"almost" completely nested.

Finally, it should be noted that a disadvantage of the closure test might be that the given system of model hypotheses has to be closed a priori. This assumption is not necessary for the other three strategies which can be applied in a much more flexible way.

V. CONCLUSIONS AND FURTHER DEVELOPMENTS

When the algorithm described in III.3 is applied, Procedure 4 should always be used, since it is more powerful than Procedure 3 and needs less computing time. A disadvantage of this procedure is that the local tests in our computer program are global Bonferroni tests, which have low power when the test statistics are highly correlated (for log-linear models, it may even occur that some are perfectly correlated). On the other hand, it is also possible that a Bonferroni-type procedure has greater power than the closure test, for example when the system of hypotheses is completely nested. Instead of global Bonferroni tests, one can try to apply more powerful local tests, for example global Simes tests or specific tests (multiple range or F tests in ANOVA models, or F resp. χ^2 tests in regression models). However, the resulting procedure may become much more complex, if the information from the elementary tests is not sufficient (as for F and χ^2 tests).

It should also be useful when the MTPs are not only performed at a fixed level α, but when adjusted p-values (see IV.1) are computed. For this purpose, modified algorithms are necessary.

Finally, it is desirable to design the input of the hypotheses and p-values more comfortable. This is possible, if the performance of the procedures can be directly based on the data. For model search problems, it is intended to provide a simpler input of the model class of interest. This also means that interactions and polynomial terms of the explanatory variables are considered.

REFERENCES

[1] Akaike H., ***A new look at the statistical model identification***, IEEE Transactions on Automation and Control **19** (1974), 716 – 723.

[2] Alt R., ***Hierarchical test problems and the closure principle***, Multiple Hypothesenprüfung - Multiple Hypotheses Testing (Bauer P., Hommel G. and Sonnemann E., eds.), Springer, Berlin/Heidelberg/New York, 1988, pp. 162 – 176.

[3] Bauer P. and Hackl P., ***Multiple testing in a set of nested hypotheses***, Statistics **18** (1987), 345 – 349.

[4] Bauer P., Pötscher B. M. and Hackl P., ***Model selection by multiple test procedures***, Statistics **19** (1988), 39 – 44.

[5] Bergmann B. and Hommel G., ***Improvements of general multiple test procedures for redundant systems of hypotheses***, Multiple Hypothesenprüfung - Multiple Hypotheses Testing (Bauer P., Hommel G. and Sonnemann E., eds.), Springer, Berlin/Heidelberg/New York, 1988, pp. 100 – 115.

[6] Bernhard G., *Computergestützte Durchführung von multiplen Testprozeduren - Algorithmen und Powervergleich*, Universität Mainz, Mainz, 1992. (Doctoral thesis.)

[7] Box G. E. P. and Hill W. J., *Discriminating among mechanistic models*, Technometrics **9** (1967), 57 – 71.

[8] Cox, D. R. and Spjøtvoll E., *On partitioning means into groups*, Scandinavian Journal of Statistics **9** (1982), 147 – 152.

[9] Dunnett C. W., *A multiple comparison procedure for comparing several treatments with a control*, Journal of American Statistical Association **50** (1955), 1096 – 1121.

[10] Dunn O. J., *Multiple comparisons among means*, Journal of American Statistical Association **56** (1961), 52 – 64.

[11] Fisher R. A., *Statistical Methods for Research Workers*, Oliver & Boyd, Edinburgh, 1925.

[12] Fisher R. A., *The Design of Experiments*, Oliver & Boyd, Edinburgh, 1935.

[13] Gabbert H. E., Meier S., Gerharz C. D. and Hommel G., *Tumor cell dissociation at the invasion front: a new progostic parameter in gastric cancer patients*, International Journal of Cancer **50** (1992), 202 – 207.

[14] Hartung J., *Statistik, Lehr- und Handbuch der angewandten Statistik*, Oldenbourg, München/Wien, 1987. (6th ed.)

[15] Hochberg Y. and Tamhane A. C., *Multiple Comparison Procedures*, J. Wiley, New York, 1987.

[16] Holm S., *A simple sequentially rejective multiple test procedure*, Scandinavian Journal of Statistics **6** (1979), 65 – 70.

[17] Hommel G., *Tests of the overall hypothesis for arbitrary dependence structures*, Biometrical Journal **25** (1983), 423 – 430.

[18] Hommel G., *Multiple test procedures for arbitrary dependence structures*, Metrika **33** (1986), 321 – 336.

[19] Hommel G., *A stagewise rejective multiple test procedure based on a modified Bonferroni test*, Biometrika **75** (1988), 383 – 386.

[20] Hunter D., *An upper bound for the probability of a union*, Journal of Applied Probability **13** (1976), 597 – 603.

[21] Kalbfleisch J. D. and Prentice R. L., *The Statistical Analysis of Failure Time Data*, J. Wiley, New York, 1980.

[22] Keuls M., *The use of the "Studentized range" in connection with an analysis of variance*, Euphytica **1** (1952), 112 – 122.

[23] Mallows C. L., *Some comments on C_p*, Technometrics **15** (1973), 661 – 675.

[24] Marcus R., Peritz E. and Gabriel K. R., *On closed testing procedures with special reference to ordered analysis of variance*, Biometrika **63** (1976), 655 – 660.

[25] Maurer W. and Mellein B., *On new multiple tests based on independent p-values and the assessment of their power*, Multiple Hypothesenprüfung - Multiple Hypotheses Testing (Bauer P., Hommel G. and Sonnemann E., eds.), Springer, Berlin/Heidelberg/New York, 1988, pp. 48 – 66.

[26] Miller R. G., *Simultaneous Statistical Inference*, Springer, New York, 1981. (2nd ed.)

[27] Newman D., *The distribution of the range in samples from a normal population, expressed in terms of an independent estimate of standard deviation*, Biometrika **31** (1939), 20 – 30.

[28] Rüger B., *Das maximale Signifikanzniveau des Tests:"Lehne H_0 ab, wenn k unter n gegebenen Tests zur Ablehnung führen"*, Metrika **25** (1978), 171 - 178.
[29] Ryan T. A., *Significance tests for multiple comparison of proportions, variances, and other statistics*, Psychological Bulletin **57** (1960), 318 - 328.
[30] Scheffé H., *A method for judging all contrasts in the analysis of variance*, Biometrika **40** (1953), 87 - 104.
[31] Schiller K., *Der Abschlu"stest zur Unterst"utzung bei der Modellauswahl loglinearer Modelle*, Multiple Hypothesenprüfung - Multiple Hypotheses Testing (Bauer P., Hommel G. and Sonnemann E., eds.), Springer, Berlin/Heidelberg/ New York, 1988, pp. 177 - 189.
[32] Shaffer J. P., *Modified sequentially rejective multiple test procedures*, Journal of American Statistical Association **81** (1986), 826 - 831.
[33] Šidák Z., *Rectangular confidence regions for the means of multivariate normal distributions*, Journal of American Statistical Association **62** (1967), 626 - 633.
[34] Simes R. J., *An improved Bonferroni procedure for multiple tests of significance*, Biometrika **73** (1986), 751 - 754.
[35] Sonnemann E., *Allgemeine Lösungen multipler Testprobleme*, EDV in Medicine and Biologie **13** (1982), 120 - 128.
[36] Tippett L. M. C., *The Methods of Statistics*, Williams & Norgate, London, 1931.
[37] Tukey J. W., *The Problem of Multiple Comparisons*, 1953. (Unpublished manuscript.)
[38] Westfall P. H. and Young S. S., *P value adjustments for multiple tests in multivariate binomial models*, Journal of American Statistical Association **84** (1989), 780 - 786.

Institut für Medizinische Statistik und Dokumentation, Universität Mainz, Langenbeckstrasse 1, W - 6500 Mainz, Germany; tel: (+49 6131) 171 32 32, fax: (+49 6131) 17 29 68, telex: 4187402 ukmdz.

E-mail: GB@DMZUK1.IMSD.UNI-MAINZ.DE

STATISTICAL APPLICATIONS OF ARTIFICIAL INTELLIGENCE

WILLIAM A. GALE

AT&T Bell Laboratories

ABSTRACT. Artificial Intelligence (AI) has now provided some effective techniques for formalization of knowledge about goals and actions. These techniques could open new areas of research to statisticians. Experimental systems designed to assist users of statistics have been constructed in experiment design, data analysis technique application, and technique selection. Knowledge formalization has also been used in experimental programs to assist statisticians in doing data analysis and in building consultation systems.

The best-explored application of AI techniques is building consultation systems. Many small systems have been built, but few systems have been offered for sale, and fewer yet have found their way into regular use. It has become apparent that this is a harder problem than was expected, although the existing successful systems suggest that it is still worth research exploration.

Analogies with successful Artificial Intelligence applications in other fields suggest other statistical applications worth exploring.

Opening new areas to research and providing new tools to users would make considerable changes in the use and production of statistical techniques. However, applying currently available AI techniques will lead to more work for statisticians, not less.

This review is an updated version of (Gale, 1987).

1990 *Mathematics Subject Classification*. 68T01, 68T20, 68T30, 68T35, 68T99.

Key words and phrases. Artificial intelligence, knowledge engineering, expert systems, statistical strategy.

I thank Marlyn Nys for her assistance with TAXSY during the Prague Summer School. Since I am not familiar with SAS, I would not have been able to demonstrate TAXSY without her. She also prepared and delivered the part of the lecture that dealt with TAXSY .

DEDICATION

This paper is dedicated to the memory of Berndt Streitberg. I first encountered Berndt in print, when he wrote "On the Nonexistence of Expert Systems." Since I was heavily engaged in the construction of expert systems then, I joined John Chambers and Daryl Pregibon in one of several commentaries on this challenging article. When I saw Berndt's reply, I felt that the differences in our viewpoints had been shrunk to the question of whether a glass should be called half full or half empty.

During the Prague Summer School, I had the pleasure to meet him. He was quite a talented and enthusiastic man, sharing his enthusiasm for Mathematica with the Summer School. He was a generous man with an understated sense of humor.

When I heard that he had suffered a fatal heart attack, I realized how much I had admired him. The world has lost a gentleman and a scholar.

1. OPENING NEW AREAS FOR RESEARCH

Knowledge Engineering (KE), the applied branch of the science of Artificial Intelligence (AI), is responsible for the *techniques and tools* used to implement the *ideas and concepts* of AI. As KE techniques are applied in various disciplines, it is becoming clear that a major contribution is new ways of building formal theories. Formalization of theories in a discipline provides a clear basis for agreement or disagreement, and thus supports combination of efforts in research. The exciting prospect from applications of AI in statistics is thus the prospect for opening new areas to systematic research.

In statistics, the knowledge that is being formalized has been called *statistical strategy*. The term covers higher level decision making than has previously not been formalized: how to translate from subject matter goals to statistical models, how to select a data analytic technique, and how to apply a technique validly.

1.2. The formal theory level.

In 1980, Allen Newell, then president of the American Association for Artificial Intelligence, suggested that AI was creating a "knowledge level." The term referred to a higher level in the well-known hierarchy of computer systems levels. While the knowledge level that he suggested does not fit well in the computer systems hierarchy, the hierarchy and its extrapolation to a higher level do provide a useful insight into the AI enterprise.

Table 1. shows a hierarchy familiar to computer scientists.

level	medium	components	composition laws
program	symbols	memory comparison arithmetic	sequential interpretation
operations	bit vector	register adder multiplexer	transfer
logic circuit	bits	and gate or gate	boolean expressions
circuit	voltage current	transistor resistor capacitor	electrical engineering
device	electrons magnetic domains	n-layer cathode	physics

Table 1. Computer Systems Levels

A level has many characteristics, of which the *medium*, the *components*, and *composition laws* are shown. The medium is processed by the components. The components provide primitive processing. The components can be combined with results predicted by a theory, the composition laws.

Each level can be defined autonomously, without reference to any other level. Programmers do not need to understand logic circuits, logic designers do not need to be electrical engineers. Yet each level can be implemented, that is *constructively* defined, in terms provided by the level below. This process is the substance of computer architecture courses.

Newell identified several features common to all these levels:

(1.) The behavior of a system defined at any level is deterministic.

(2.) Total system behavior results from local processing by each component.
(3.) Immense variety of behavior is obtained by an immense variety of ways of assembling relatively few types of components.
(4.) The medium is realized through stable states of matter, such as electrons in a capacitor.

He also pointed out that each level is implemented through a great *restriction* on the components which can be constructed using the means provided by the level below it.

Newell then proposed a "knowledge" level above these levels. As he discussed, his proposed level violated all four of the points noted in the previous paragraph. There may be several levels above the program level, and a knowledge level may be one of them. But the most fruitful next level would seem to be one that extended these same properties which have worked so well at lower levels. Much of current AI work can be seen as work towards a level above the programming level that *does* continue all the properties noted above.

The level under construction might be called a "formal theory" level. This level is *not* complete, and is not as well defined as the lower levels, but in its current status it has begun to be useful. The medium of the formal theory is the statement, a sequence of symbols well formed according to some syntax. The composition rules are logics, such as first order predicate calculus or modal logic. The components most suitable for this level are not yet clear (or perhaps this level will give up this property). In their place we currently find a set of *concepts* and a set of *techniques*. The concepts include belief, knowledge, actions, goals, problems, and reasons. The techniques include rules, frames, and objects.

A number of formal theories, of varying degrees of completeness, have been built using programs as a medium of implementation. The examples given later show what has been done in statistics. Since the components for building such theories are not yet standardized, current practice is to focus on the concepts, and to use the techniques available as seems most appropriate.

The contribution of AI research is KE techniques and tools for building formal theories. Formal theories have long been built using mathematical tools, and the progress made using them suggests why new tools for formalization are exciting. What distinguishes current AI programming is the attempt to build programs that formalize such concepts as goals, problems, and actions. These concepts occur in statistics, as in any rational activity, and in data analysis have come to be called statistical strategy.

1.2. Opportunities for formal theories in data analysis.

Two rather similar views of the data analysis process have been proposed by Hand (1986a) and Oldford and Peters (1986a). Hand discussed four *stages* of analysis, while Oldford and Peters distinguished four *levels* of strategy. That is, Hand was concerned with entities which actually take place at different times, while Oldford and Peters' description is more of a classification. Still the views are similar, and a comparison may give some feeling for what in data analysis needs to be represented.

Hand's four stages are (1) formulate aims, (2) translate into formal terms, (3) numerical processing, (4) interpretation. These stages were given specifically as stages in a multiple analysis of variance (manova). The first stage is concerned with what dependent and independent variables are involved, how they are related, and what questions the researcher wants to explore. It occupies a large part of the time in actual consultations. The second stage results in the translation from a problem statement in the ground discipline to a problem statement in statistics terms. The third stage consists of estimation, testing, data cleaning, and transformation. This stage functions within the statistician's language. The fourth stage consists of translating back to the ground domain. As Hand points out, there will be various loops in an actual analysis, returning to earlier stages to alter decisions. While given as stages in manova, I believe they present one reasonable view of data analysis.

Oldford and Peters suggest 'operational level' as a scale for thinking about procedures. They illustrate the idea rather than define it, but it seems to be related to a possible hierarchical organization of procedures. At the lowest level are standard numerical procedures of statistics, such as least squares fitting or robust fitting. Selections from this level constitute the minimal components of a statistical package. Just above this level are such sub-procedures as collinearity analysis and influential data diagnosis. Each of these presupposes the existence of procedures in the layer below it. Above this layer lies a layer of techniques, such as regression analysis, spectrum analysis, or analysis of variance. The top-most identifiable level has strategies for analysis and for design. This is another reasonable view of data analysis.

The levels idea rests on a notion of a procedure using other procedures as building blocks to carry out its goals. The notion of stages is that of what is done first. The relationship between them is that the high level strategies are used first and more frequently. The low level strategies are used later if at all. Thus the higher levels of a hierarchy of techniques will correspond to the preliminary stages of a study.

Wittkowski (1988) sketches a program that explicitly recognizes six different levels of abstraction, analogous to those discussed by Hand, Oldford, and Peters. The implementation realistically confines itself to the lower levels of abstraction.

1.3. The Reality of AI Applications in Statistics.

Streitberg's article "On the Non-existence of Expert Systems – Critical Remarks on Artificial Intelligence in Statistics" (1988) forcefully raised the issue of how much progress was being made toward commercially available systems, and how knowledge-based systems could be tested. Wittkowski, (1990), reminds us of the same fundamental problems. Streitberg argued that there were no commercially available systems, and that therefore there were no expert systems. At about the same time, I wrote (1988):

The current status of consulting systems is that several feasibility systems have been described in the literature. None have become prototypes, that is, systems used by a few friendly users. There are no commercial systems based on formal models of data analysis.

Thus we agreed on the non-existence of commercial systems. However, there are at least two discernible stages of development that come before this stage. A feasibility demonstration is a program that the original designer can demonstrate to others. A prototype is a program that people other than the original designer can run. The extent of my disagreement with Streitberg was over whether to call such discernable stages of research expert systems, or not. This is akin to whether one person will call a glass half full of water that another calls not full.

I also wrote in 1988:

In the next five years, I expect to see perhaps a few dozen research systems, several prototypes, and probably one or more commercial systems.

In the remainder of this review, I will show that this was an accurate forecast.

One sign of maturity as systems become more available has been an increased discussion of critical parts or processes in the construction of consultation systems. An awareness of the importance of the interface to the user is one such critical part. Nelder (1988) describes the interface for GLIMPSE, a program discussed below. His most important conclusion is that the interface is at least as important as the expertise. Hietala (1992) describes her use of hypertext, a uniquely computer oriented format, for explanation in her system ESTES. The proposal for AMIA (Ollivier and others, 1992) proposes to provide a front end to an existing energy demand forecasting program. As such, interface issues are a primary concern.

A related study is an empirical study of how consultees relate to consultants by Clayden and Croft (1990). The study reached the negative conclusion that machines would not be able to mimic the current process. This result should probably have been expected, since in many "intelligent" programs, such as chess playing programs, progress became possible when attempts to model how people do the task were set aside and the different strengths of the machine were used.

Pregibon (1986) gave a number of heuristic suggestions based on his experience in constructing the strategy for REX. Herrmann and others (1991), have given a discussion of knowledge acquisition for a front end that was developed for the PROPHET system. PROPHET is an American national resource for biostatisticians used by about 1000 researchers in 40 institutions. The approach used was far more organized than the informal process described by Pregibon. In particular, it took as a starting point a published paper strategy, that of R. G. Miller's *Beyond ANOVA*, (1986). This provided a minimal hierarchical network of decision trees at different levels of abstraction. As usual, when an initial strategy is formulated, many holes were detected. However, the original framework was sufficient to permit an orderly expansion of information.

Van den Berg and Visser (1990) have studied the controversial issue of how much agreement there is among statisticians on strategy issues. It has generally been supposed, and sometimes remarked informally (Bell and Watts, 1988), that statisticians differ considerably in how they approach the same problems. Given this, the amount of agreement found by Van den Berg and Visser is surprising, and quite possibly useful. Considerable agreement between the statisticians was found in classifying methods for exploratory or confirmatory analysis, and in classifying methods as suitable for discrete or continuous variables, and in grouping methods based on generalized linear models. However, there was no agreement on the classification of methods not based on generalized linear models. A principal components analysis showed a first dimension of methods with dependent variables versus those without, and a second dimension of methods for one or two variables versus multiple variables. Van den Berg and Visser's results are controversial because they show both agreement and disagreement. The procedure used was an open ended classification test, which would tend to accentuate differences. While not all of the tested statisticians drew the distinction between methods with and without dependent variables, for instance, I would expect perfect agreement on a specific sorting task asking the statisticians to make this distinction. While this study remains the chief data, there will undoubtedly be discussions of whether the glass is partly empty or partly full. For me it is partly full.

1.4. Organization of the Paper.

In the next section I discuss several computer programs that explicitly formalize statistical strategy under headings based on a combination of the Oldford and Hand views. The first heading is translating a research goal into a specific data analytic agenda. This level is represented by a program, RX (Blum, 1982), that takes a research question posed in medical terms and produces a description of the statistical study that needs to be done to answer the question. This level corresponds most closely to Hand's 'translate into formal terms'. It is the first point that statistical knowledge enters the study. It is characterized by need for knowledge of both a ground domain and statistics. This is an old program now, and the lack of progress in this area suggests that it is particularly difficult.

At this point, we know that some statistical analysis needs to be done. The second heading is choice of technique, which assumes that some analysis needs to be done, but that a technique has not been selected. The program MUSE (Dambroise, 1987) remains the best developed at this level, although other programs have made some advances. This level is characterized by a lack of assumptions or restrictions on the study, leading to a small role for formal statistical calculations, and a large dependence on information which cannot be gathered by examining the data.

The third heading is analysis given the technique, is now represented by a multitude of programs, including MULREG, a commercially available and widely used product. This level is characterized by active use of statistical tests, plots, and transformations to detect violated assumptions and take corrective action.

The following table summarizes the above discussion of the levels of statistical strategy used to organize this paper. The key problems included in the table are discussed later.

The following table summarizes the above discussion of the levels of statistical strategy used to organize this paper. The key problems included in the table are discussed later.

Level	*Characteristic*	*Key Problems*	*Example*
formalization	translation into statistics	represent knowledge in two domains	RX
technique selection	need information not in data	get information from naive user	MUSE
technique	test many assumptions make corrections	represent all assumptions corrections uniformly, deal with interacting violations, order of corrections	MULREG

Table 2. Levels of Statistical Strategy.

2. EXAMPLES OF AI APPLICATIONS IN STATISTICS

This section reviews the examples of AI applications in statistics that I now consider to be the most important. Other reviews of work in this area have been published by Hahn (1985), Gale (1986e), Hand (1986b), and Chowdhury (1987).

2.1. Translating Research Goals to Statistics.

Translating a research goal into a specific data analytic agenda is a high level and difficult task. The one system built at this level shows that some progress can be made with current techniques, but suggests also the difficulty of an open ended system at this level.

The system is RX, built by Robert Blum (1982). It was intended to discover causal relationships in medicine automatically. As developed, it was limited to relationships derivable from a single data base. This limitation allowed substantial research on automatic study design, but suggests that considerable generalization remains to be done.

The data base RX used was a subset of the American Rheumatism Association Medical Information System rheumatology data base. This data base was developed for doctors to record symptoms, lab values, and therapies of patients seen in seventeen rheumatology clinics. RX used the subset of records consisting of the most active 500 patients with lupus erythematosus at the Stanford Rheumatology Clinic. The patients chosen averaged 50 visits over four to eight year periods. These patients are not a random sample, but a convenient one to build a feasibility study on.

These patients were seriously ill, with multiple diseases and multiple therapies. The number of concurrent diseases and therapies makes the determination of causal relations challenging.

RX was intended to emulate the scientist at work, with three major modules: a discovery module to suggest relationships to study, a study module to refine the suggested relationship to a data analysis problem, and a statistical module to carry out the data analysis required. All three used the knowledge base which is described below. The discovery module and the data analysis module were only minimally developed.

The data analysis module, for instance, simply performed what statisticians call an Ordinary Least Squares regression rather than doing a complete analysis. The module is given a dependent variable and several explanatory variables. Good statistical practice considers the choice of a model (such as transformations applied to the various variables) and the choice of method (including weighted least squares, maximum likelihood, and many others). This module always uses unweighted least squares regression on the untransformed variables. Therefore this module represents essentially no statistical knowledge.

The study module, which tackled the problems posed by the multiplicity of diseases and therapies, was developed to the feasibility demonstration stage. In performing its task, the study module uses three main structures for its knowledge base, representing medical concepts, causal relations between the concepts, and the study design.

RX represents medical and statistical concepts in classification trees with inheritance. For instance, prednisone is a steroid, a steroid is a drug, a drug is an action; cholesterol is a chemistry, a chemistry is a lab value, a lab value is a state. A few hundred medical concepts were included in RX. All are categorized as a state of the patient or an action that can be taken. In the statistics tree, the terminal nodes represented the few tens of methods available to RX by calling on the IDL statistics package. However, it appears that this portion of the knowledge base was not widely explored, since regression was the only technique used.

The causal hypotheses portion of the knowledge base was designed to be the portion that would grow as the automated scientific process was carried out. To start with, it was a representation of relevant results from the medical literature. The knowledge was formalized as a labeled directed graph. In the graph, causes and effects from the medical concepts tree were nodes, causal effects were the arcs. The arcs were annotated with considerable information about the relationship. The nodes and some arcs within this network were to be specified by users of the system. The discovery module would then suggest which additional arcs appeared the most interesting to study. After data analysis, the information in a studied arc would be filled in.

The information included in the arcs represented characteristics of the causal relationship such as setting, intensity, and frequency. The setting is an arbitrary boolean predicate representing the conditions under which it is known that a causal relation exists. Typically, the relationships studied by RX were studied by longitudinal regression models on some or all of the patients in the sample. The distribution across patients of the regression coefficients for the causal variable is stored to show the frequency and intensity with which the causal relation holds. The average regression coefficient is compared to a scale of importance previously input for each effect to measure the importance of the effect. While information on whether the effect is an increase or a decrease is implicit in the signs of the coefficients, it is explicitly represented also. The functional form of the transformation used, including time delays, completes the information summarizing results of a study from the data base.

The arcs also include representations for validity and evidence. These refer to the state of proof of the relationship rather than its apparent strength. The validity of a causal relation is considered to be highest if has been repeatedly confirmed in prospective randomized studies. A single study (such as done by RX) could rank midway at best on the scale used to rate validity. Evidence includes literature citations for relations from published literature, or a list of patients used for a study by RX.

In performing its tasks, the study module fills out a frame of information to pass on to the analysis module. The contents of this frame represent important information for the design of an experimental test. The selection of which things to include in the frame represents important knowledge included in the program by its designer. The first step is to parse the hypothesis and verify that each concept is operationally defined in terms of the underlying data base. The parse is stored. The second step is to identify confounding variables. These are any variables that might effect either of the two principal variables of the hypothesis. These are determined from the causal network and stored. A method for controlling confounders is then selected from (1) dropping patient records, (2) eliminating affected time intervals, (3) incorporation in multiple regression. This choice is made using rules.

Production rules are also used to choose between cross-sectional and longitudinal designs. The production rules used are an additional knowledge representation technique in RX, but not, it appears, a major one. The statistical method is selected, and the database access calls are constructed and stored.

A key issue at this level of strategy is representing knowledge from two domains, the *ground* domain and statistics. The ground domain is the domain in which statistics is being applied, a small part of rheumatology in the case of RX. RX shows that for a selected ground domain, knowledge from two domains can be represented and used. However, it has become apparent that *scale* is an important part of an intelligent system, and RX does not show that a small part of medical knowledge can be scaled up significantly. Since statisticians are used to computer systems that do not depend on which ground domain they are applied in, they may find this limiting. However, the state of the art in knowledge representation will not support knowledge in all domains to which statistics might be applied. The best that can be hoped for in the near future is a system with all the requisite statistical knowledge, to which ground domain knowledge for one domain can be added by a ground domain expert.

The RX system is a state of the art AI system, and deserves study to show current limits and capabilities. It appears that within the limits of the concepts in a single data base, those concepts could be defined sufficiently well to permit design of experiments. However, the resulting system would need considerable human input to deal with additional data bases. Dealing directly with humans in experiment planning would be even more difficult. The causal network established would seem to be useful for other programs, perhaps a clinical consultant.

2.2. Selection of Data Analytic Technique.

RX, just discussed, has an automation of method selection, based on its statistical technique tree. Each node of the tree has represented objectives, prerequisites, and assumptions. Prerequisites are properties that must hold for mechanical applicability of the technique. Assumptions are properties that must hold for validity of the results. Selection of technique is made by matching the study requirements to the goals and prerequisites. The program was not developed so far as to use the assumptions. Also, since regression was always chosen by RX, this portion of the knowledge base was not well tested. For a study that was not formulated by a machine, RX does not have suggestions of how to acquire the description of the study to match against technique descriptions.

STATPATH was described by Portier and Lai (1983). The program used carefully worded questions to perform a binary tree search of the techniques known to it. This structure was represented by a set of production rules. Thus STATPATH proposed to determine the crucial aspects of the description of the study by asking the user. For several years, this system represented the best approach to selection of a data analytic technique.

However, the key issue is that when the user does not know enough statistics to make a technique choice, he may well not know enough statistics to answer questions well. Portier and Lai were aware of this issue, writing "how the question is asked determines the validity of the answer." Besides careful wording of questions, they provided several other help mechanisms. Additional information on the question could be gotten by replying "?" instead of "yes" or "no". Information was provided on each procedure the system knew about, and this information could be browsed. The user could reply "unknown" to any question and get *both* further lines of questioning.

While these approaches all look useful, it is not clear that they are sufficient. Indeed, Portier and Lai reported "we have spent considerable time on the wording of statements and questions and we still feel that we do not have statements which would be completely understandable to the non-statistician." The exploration of multiple branches may be useful when only a few questions are not understood, but the techniques to consider increase exponentially in the fraction of questions not understood. None of these techniques address the problem of *mis*understanding – the user thinks he understands, but does not. It will be difficult to give the user confidence that he is using a good technique, and that he can defend the choice of that technique.

SAM (Athey, 1989) uses a similar technology, however Athey did not stop at evaluating the program based on its structure, but tested its effectiveness in use. The subjects were students finishing an introductory statistics course. Students were randomly assigned to one of two test groups, one which would use SAM, and another which would use a traditional text book. None of the students in the expert system group had ever used the computerized system before, and none had received any previous instruction in the search process used by SAM. The text book group used the same methods that they had been taught and had been using for two semesters. All subjects had the same five types of problems covered in their introductory statistics course.

The students using SAM performed significantly better finding 38/60 correct method selections, while the textbook group scored 27/65. There is an important lesson to be learned from this. Simple methods can effective. We do not need to use the most advanced techniques to make useful programs.

Another example of testing a choice guidance program is given by Molenaar and Passchier (1992), who describe and test an advice giving program to accompany or to be embedded into the AGREE program. The AGREE program calculates thirty some indices of inter-observer agreement. The AGREE CONSULTANT uses one rule per index to assist users in making the correct choice of index. The program was tested with sixteen subjects without knowledge of either the AGREE program nor the agreement coefficient literature. A correct recommendation was obtained in 395 out of 416 subject-tasks, certainly a good performance, although there seems to have been no control group using just AGREE.

The experiment tested four variations of the CONSULTANT, and may have been more concerned with guiding the development of the advisor than with validating its utility.

"A Statistics Advisor" (ASA) by Richard O'Keefe (1985) provides a sharp contrast in approach. ASA formalizes knowledge about measurements – their sorts, control, and sequence – and about the relationship between measurements and useful analyses. While ASA also did not address the key problem of interacting with a statistically naive user, it shows some possibilities that might be developed in the future. The possibility it opens is that of communicating with the user about data gathering protocols, which could be done in the language of the users specialty.

ASA formalizes descriptions of value spaces and experiment structures. O'Keefe first tried using a traditional classification of value spaces familiar to most statisticians: nominal, partially ordered, ordinal, interval, and ratio scales. He found as he began to try using this classification in a formal setting that it had several deficiencies.

For instance, counts do not fit into this scheme. They are stronger than a ratio scale in one regard, that their unit is fixed, and yet they are not closed under all operations that can be applied to a ratio scale. The classification fails to distinguish strictly positive scales from differences of such scales. That this was an important distinction became clear when he considered rules for transformations of variables. The classification fails to distinguish linear scales from periodic ones. It does not include permutations. It fails to distinguish the unbounded from the practically bounded (such as lifetimes or heights) from the formally bounded (such as percents). O'Keefe thus found (in analogue with most formal theory creation) that he needed to set up a new method of classification from which the important distinctions could be derived.

A flavor of the lattice representation that he built can be gotten from the following simple example. Counts of oranges and counts of apples are two different value spaces, neither of which subsumes the other. It is still possible for ASA to add 3 oranges + 2 apples, with the result being 5 fruit, since fruit is the common supertype of both apples and oranges in the object taxonomy. The concepts represented in ASA's value spaces include classifications (such as the object classification), approximations, counts, locations, physical dimensions (such as time, length, and mass), directions, and arithmetically derived scales (which include sum, difference, product, ratio, and proportion).

An example will illustrate the importance of a description of a value space in statistical applications. Velleman and Hoaglin (1981) give an example of describing a univariate set of data representing rainfall. The value space for the data is *length(water)* length(water)* length(water)*. Three rules relevant to this value space are: 1. Always consider the raw data. 2. Consider logs if the value space is positive. 3. Consider the kth root if the value space is a kth power. The third of these suggestions leads to a reasonably symmetric histogram. It is important to note that these rules were not created for this example, but are general rules applying to all transformation goals, and were first added to ASA to handle other examples.

In the text a data based "transformational shotgun" is applied to this example. A transformational shotgun is a brute force statistical technique which just applies several transformational powers to the data, usually the square root, the log, the inverse square root, and the inverse. In this case the shotgun showed that neither a square root nor a logarithm was entirely satisfactory. The point is that a useful transform is suggested by the type and not by the data.

O'Keefe went on to give some primitives and a syntax for a language to describe experimental structures. He described the primitive steps of an experiment in terms of a triple giving the description of the unit operated on, the variables known before the step and the variables known after the step. Primitive steps that he used included: the identity step (after variables equal before variables), an observation step (after variables equals union of before variables and observation variables, *with the order of the observations guaranteed to be immaterial*), and the vague step (after variables equal union of before variables, treatment variables, and observation variables, *with nothing known about relationships*). Several ways of combining primitive steps were identified, including sequential combination, parallel processing of components of the original unit, parallel processing of randomly assigned units, and parallel processing of randomly selected units.

The principle contribution of O'Keefe's work is the formal experimental description language. It can be used to select an appropriate method for analysis of the collected data, and to make a statement about the generality of the resulting conclusions. It will still take considerable work to make practical systems using this approach. The range of descriptions for which ASA could determine analysis methods was certainly not trivial, but just how big it was is not clear. A larger problem was that ASA had no method of building a description by interviewing the user. As I commented above, getting sufficient information from the user to carry out an automated task is a key problem.

MUSE, by Eric Dambroise (1987), was another advanced system proposed for determining the choice of a technique. MUSE was intended for use at a single industrial location, but by all the departments – personnel, engineering, accounting, etc. – at that location. MUSE was a doctoral project, reached the stage of a feasibility demonstration, and was not developed further after the thesis.

MUSE has three modules that together select the analysis methods used. First, the user indicates which data is to be used. The system then uses rules and queries to the user to classify the data. Some of the "data" may actually be identifiers. Otherwise a part of the data may by continuous, ordinal, nominal, or dichotomous. Building further, data may be recognized as a higher level construct, such as a matrix of similarities or dissimilarities, or a contingency table. It appears that much of this can be decided from examining the data, and that the rest can be determined unambiguously from the user. This classification of data is less well formalized than is O'Keefe's classification, but it has been shown that the information required can be transferred to the machine by the intended users.

Second, the objectives module selects a set of elementary objectives required to provide service to the user. To do this, the module formalizes the set of statistical tasks that are performed in the industrial location in terms of elementary objectives available through the known statistical techniques. The statistical tasks are represented as "scripts" (a KE technique for representation of frequently occurring sequences of events) of the elementary objectives. It is easy to add scripts. The module uses meta-rules and plausibility factors to select a set of scripts for final selection by the user. The system provides some help to the user in building new scripts if none of the existing ones are satisfactory.

Third, the methods of analysis are selected. This module has a list of available methods, each characterized as to the inputs required and the elementary outputs provided. The statistical technique is selected by matching the description of the inputs available (prepared by the data module) and the description of the outputs required (prepared by the objectives module) to the list of available techniques. This module is the first functional implementation of technique selection based on a formalization of the inputs and outputs of techniques. This method was sketched by RX, but RX never selected any technique except regression, so it did not test the ideas. Thus the data, objectives, and technique selection modules in MUSE provide a substantial advance in automatic technique selection.

MUSE was the first system to demonstrate a means of getting from the information that a statistically naive user can competently provide to a choice of technique. It accomplished this by formalizing the relationship between what the user does know (attributes of data and task to be done) and what the user does not know (inputs required and provided by various statistical techniques). The formalization was achieved in part by taking a narrow view of the analysis, and the approach needs to be tried in a more comprehensive setting.

2.3. Application of Data Analytic Techniques.

After translating a domain problem into statistical terms, and after selecting an appropriate statistical method, that method needs to be applied. But application is not just a calculation. Application requires analysis, because peculiarities of the data may violate some assumptions of the selected method. The last five years have seen substantial progress in this area.

The first system to demonstrate using AI techniques to automate the application of a statistical method was REX. The REX system (Regression EXpert), was built by Daryl Pregibon and me. Gale (1986a) gives the most extensive description. REX advises a user in the analysis of regression problems. It guides the analysis by testing assumptions of regression, suggesting transformations when assumptions are violated, and justifying its suggestions when requested. It interprets intermediate and final results, and instructs the user in statistical concepts.

As REX begins, the user sees several windows on a bitmap display terminal, one for dialogue, one for plots, others for various information. The users must provide REX with a set of measurements of two variables, and must know that they want to do regression analysis. If a system at this level is not combined with assistance in selecting a technique, these requirements will set the skill level required to use it.

The first interactions in REX are personalization questions, such as how thorough the analysis should be. The session continues as REX checks for assumption violations. REX provides information on the interpretation of tests as it runs. At points that the user must make a decision, REX offers to show relevant graphs. REX is always prepared to define terms through a lexicon, explain what a test does, or state why it is suggesting a particular transformation. REX concludes the analysis by issuing a four page English report on the data and the results.

The strategy REX uses initially accepts the data as given, and assumes a linear model together with ordinary least squares as the fitting method. REX then checks these assumptions in detail. It first checks for superficial problems in any one variable, then it checks each independent variable for linearity, and finally the residuals are checked. At any point that a problem (a violated assumption) is found, REX considers possible transformations of the data, the model, or the fitting method to alleviate the problem. If a transformation will solve a problem, REX suggests the transformation to the user. REX terminates either by solving all problems or by locating a problem for which it cannot find an effective and acceptable solution.

The regression strategy used in REX handled a wide variety of problems in actual data sets. It was not as fully developed as it would need to be for a product, but the techniques used would suffice to extend the strategy to handle all commonly occurring problems. There would remain some uncommon problems that it would not be worth the effort to have REX handle.

REX made two major contributions to subsequent work. The first is a viewpoint for thinking about data analysis as a diagnostic problem. Briefly, one should list model assumptions (analogous to possible diseases), test the data set at hand for violations of the assumptions (analogous to symptoms), and if found select a transform of the data (analogous to treatment). The success of this approach depends on the representation of statistical knowledge. This is the second major contribution of REX. REX has a set of statistical primitives including tests, plots, assumptions, and transforms, which can be implemented as frames with slots containing procedures, or as objects (classes) with instance variables and methods. The hierarchical structure of the network of frames directs the interpretation of the statistical knowledge. The classes of frames used in REX provided us with an initial list of classes of primitives that has remained useful and has been expanded. This conceptual model provided a key input for work on Student, as described below.

Thus, we developed a means of representing strategy, providing an initial means of formalization. We found that the most convincing explanations in statistics were provided by graphs. A report in English and graphs provides a useful permanent record. Definitions are easy to provide. Verbal explanations of why a transform was suggested were possible, but not as convincing as graphs. Based on finding means to solve these two programming challenges, we have called REX a feasibility demonstration. It demonstrates the feasibility of using direct construction with artificial intelligence techniques to provide statistical expertise.

Since the experience in constructing REX was described, several other demonstration systems have been built. These include MUSE, Express (Carlsen and Heuch, 1986), RAO (Drapier, 1987), and unnamed systems by Berzuini and others (1986) and by Darius (1986). These systems intend to model the actions of a human analyst, as did REX.

REX never progressed beyond the feasibility demonstration stage, in part because it was built on a Vax which would not be widely available. With the widespread acceptance of the IBM PC and clones in the years since then, many systems have been built for these readily available machines. An example is THESEUS briefly described in (Bell and Watts, 1988) and more extensively in Bell (undated). THESEUS was a thesis project for Bell, and work seems to have stopped with her graduation. However, I was able to get a copy of the program and to demonstrate it at the Prague Summer School. Thus THESEUS is one of several prototypes that have now been built.

THESEUS is a shell consisting of a rule editor and a rule interpreter. It was written in Pascal. Bell developed a set of rules concentrating on the area of one-way analysis of variance and related techniques, because preliminary research showed these were heavily used in industry. The intended users were industrial statisticians, and she had the cooperation of half a dozen of her intended users while she developed the system. When she found that each of her statisticians approached a problem in a unique way, she decided that the rule editor must be an important part of the design so that the users could revise rules to fit their own approaches.

THESEUS is a production rule system with two types of rules. The first type deals with the selection of an appropriate technique using forward chaining. The second type verifies the applicability of tests by checking assumptions, possibly making transformations, and possibly removing outliers. These rules are treated as backchaining rules. These operations are quite analagous to those that REX undertook. The advances represented by THESEUS are the convenient rule editor, a smooth interface on a popular machine, and active testing of the software by prospective users. THESEUS does not address the problem of connecting the advisory system to a preexisting and tested statistical package; the routines needed for the application were also coded in Pascal.

Dorda, Froeschl, and Grossman (1990) describe another successful prototype, WAMASTEX. This program has focused on exploratory data analysis, favoring simple univariate techniques. The program is intended for use by clinical physicians, and has been used in the Vienna University Hospital. The program guides the user through data validation, scale determination, analysis of outliers skewness and modality, selection of an appropriate set of descriptive measures, and output of results graphically, numerically, and linguistically. The heuristics used in the program were tested experimentally by getting responses from 37 practicing statisticians.

WAMASTEX addresses the problem of connecting the logic to an existing statistical system by building the system within SAS(R) using macros. This is probably sufficient for its simple techniques, but SAS was not designed for symbolic programming, and scaling the work up might be difficult.

The usage and acceptance of WAMASTEX was monitored by appending a questionnaire to the consultation system. The authors report, based on 100 sessions of usage by 60 users, returning 32 inquiry forms,

Notably dialogue management and system usage have gotten high scores throughout. The statistical results have been judged as correct, sound and presented in an easily comprehensible layout.

Another student project that has reached the prototype stage is Hietala's ESTES (1988). The system is in use at her school, but since it was programmed for a Macintosh, I was unable to demonstrate it at the Prague Summer School.

Lubinsky explored another approach in *TESS* (Tree Environment for Statistical Strategy), which reached the stage of feasibility demonstration. An application of TESS to regression was described by Lubinsky and Pregibon (1987), while Gale and Lubinsky (1986) described the knowledge representation used by TESS. Rather than try to capture the intuition of human experts, TESS aimed to substitute the machine's ability to compute rapidly.

The knowledge representation used in TESS is again a tree of features and transforms, and each feature or transform is represented by a frame. While REX uses such a representation as a procedural guide, TESS uses it as a representation of a space of descriptions to explore.

TESS seeks descriptions that are both interesting and accurate. There is a trade off between these two, since the most accurate description of a set of data is the data set itself, but this is also the least interesting description. Depth in the tree of features is the measure of interest, because with a hierarchical tree the lower nodes are more specific. Accuracy is estimated by techniques specific to each node, expressed as a number between zero and one. There is no theory for combining accuracy and interest measures, but the resulting values for several descriptions can be plotted to facilitate comparison.

TESS's search strategy is a modified depth first search. The objective is to deal in complete descriptions (which is achieved by reaching a leaf of the tree), but to explore differences likely to be important quickly (which is achieved by branching near the root of the tree). The heuristic search algorithm thus produces a set of descriptions which the user can judge for interest and accuracy.

GLIMPSE, was a larger scale project than the demonstration systems described so far and reached the stage of being offered for sale. At the price set, however, sales were minimal, and it was withdrawn. GLIMPSE was designed as a front end for the widely used General LInear Modeling package, GLIM, and was described by Wolstenholme and Nelder (1986). GLIMPSE was built with sigma-Prolog and Augmented Prolog for Expert Systems (APES) on a Vax(Tm. The purpose was to make GLIM accessible to more users.

The work on GLIMPSE does suggest a way to solve two important and related engineering problems. The first of these is how to make a system that a naive user can learn from, slowly becoming an expert user, without the system impeding expert users. One way this might be accomplished would be to have the user always in control of an interactive system, but with the system having a sufficient model of the analysis to assist at any time. While this is another unsolved problem, GLIMPSE suggests a way to approach it.

GLIMPSE provides a task command language to the statistician, and three levels of assistance in using it. A single question mark entered while giving a command is a request for a *reminder*. The system replies with the syntax of the command that has been started. A double question mark is a request for *prompting*. GLIMPSE then takes a more active role, providing keyword options as menus, and asking questions (with explanations available) to determine parameter values. Completely specified commands are shown to the user for approval. On request, GLIMPSE will enter a *hand holding mode* for a specific task. In this mode, the system will suggest a complete command to use. It is this capability, to provide suggestions at *any* point in the analysis that is new.

The reason it is difficult is not too hard to see: when *I* watch a statistician using an interactive statistical system, I am not always sure what the statistician is up to. Inferring intentions from actions is a difficult problem. The way REX resolved this problem of how the machine would have a model of what was being done was to maintain control. But this way does not provide a system that moves gracefully from novice to expert.

The approach in GLIMPSE is two fold. First, there is a formal model of the analytic process as composed of nine *activities*. These activities include determining attributes of the data, as in MUSE, model selection, and model checking. Only some of the transitions between activities are allowed. The activity determines what the program assumes the statistician's intention is. Within the activity, then, each command that the statistician can use not only does the requested activity *as in a conventional statistical system* but also builds a formal structure representing what has been done. This structure is the basis for the machine's suggestion if it is requested.

Part of the model selection activity was described by the Wolstenholme and Nelder (selection of link and error were not discussed). The activity will select a subset of variables to use from those available. The user specifies an initial kernel of variables that should be included based on prior knowledge. The remaining variables are free terms that may be added to the kernel or may be dropped. From any given model, the statistician can form a new model by dropping some of the free terms or by moving some free terms to the kernel, or both. Models with cross terms can be created. For a given model, the system will provide statistics for adding any given free variable to the kernel and for dropping it from the complete model with all kernel and free variables. These statistics are numbers which increase with the explanatory power of the variable added or deleted, and which have well known levels when a purely random variable is added or deleted.

The structure maintained by the system to represent this search is a graph whose nodes represent models. When asked for advice, the machine uses rules based on the statistics for proposing additions to the kernel and deletions from the free terms. Some of the rules may provide more than one suggestion. A graph of suggestions is then generated, and is explored depth first.

A novice relying entirely on the help of the system will make a mechanical search, which, indeed, the machine could do by itself. As experience is gained, the user may learn some shortcuts, and this is the reason for pulling a novice through the search. If experts normally select a model in this way, they will be glad to have the bookkeeping assistance provided. If they think in some entirely different way, such as Mallows' (1973) *Csubp* statistic, they may be frustrated. (*Csubp* is a statistic that includes both the residual sum of squares, which will decrease as more explanatory variables are introduced, and a penalty for the expected bias from having too many explanatory variables. One normally uses it by examining a plot of many, or all, possible models simultaneously. Its use is thus not compatible with a stepwise approach.)

The activity module thus provides its users with the same primitives as are required for a mechanical approach to model selection. If this formalization of the model selection process is unduly restrictive for the expert, or unduly tedious for the novice, then the approach may not be successful.

GLIMPSE is an example of what seems to be a frequently successful approach: building a knowledge-based front end for an existing statistical package. Duijens, Delft, and van den Berg (1988) describe PRINCE, a program that prepared input statements for an older program with arcane input requirements. Prat and others (1992) have been developing tools for the construction of front-ends, and have used the tools themselves to build a few small front-end programs.

The most commercially successful application of AI concepts in a statistical program has been MULREG, described by Dumouchel (1990). MULREG was offered as part of an upgrade of an existing statistical package sold by Bolt, Baranek, and Newman, and was bought by all of their installed customer base, over 500 industrial sites.

MULREG addresses linear models for multiple variables, with support for stepwise selection of variables, robust estimates, residual diagnostics, transformation and back-transformation of variables, checks of distributional assumptions, automatic treatment of categorical variables, unbalanced designs, and various other processes. The structure of MULREG can be likened more to a sophisticated word processor than to the typical expert system, since there is an "analysis object" which persists between sessions and can be created, named, copied, and edited using MULREG.

This analysis object has considerable internal structure. The major categories it represents are data, models, and output specifications. Data distinguishes variables, such as temperature or pressure, and their descriptions, from the terms used in a model such as log(pressure) or pressure *times* temperature, and their descriptions. Multiple models may be incorporated, with a sub-object representing each model. The model object records which variables it uses, excluded cases, weights that are used and so forth for the model as a whole. It also records for each response variable it name, transformation, coefficients, residuals, Anova information, and influential points. The output specifications include graphical specifications, technical choices, and optimization specifications.

MULREG's strategy, expressed as the set of techniques it makes available to the user, weighed several factors. Where possible, commonly accepted techniques are presented. However, the multiplicity of options led to using some techniques that were general enough to cover many options rather than a special technique that would be best for some particular set of options. As is common in developing strategies, MULREG turned up some combinations of options for which no technique had been published; in these cases, standard techniques were extended in as straightforward a way as possible.

In particular, MULREG offers a graphical interpretation of the data at each decision point, along with a verbal description of decision to be made and the choices possible. The use of graphs has been found to be a powerful communication method and is offered in many statistical systems. Note that a statistical system, even if very intelligent, is probably not yet able to assess whether a communication method is more powerful than another. In a number of cases, the developers needed to construct novel graphical displays.

Among the possible choices that a person can make at any time, one or more is highlighted as suggested. This provides guidance to new users while allowing other choices if the user wants.

Other commercial systems are appearing. However, the two described by Raes (1992) are exceedingly simple. They do provide formalizations of statistical strategy, but the simplicity of the formalizations raises questions about their accuracy. Raes's paper is interesting in that it gives the complete strategies embedded in the systems, so that they can be examined by others. Their simplicity, aside from any questions of accuracy, also suggests that their duplication might be a useful student exercise in a course.

2.4. Assisting an Expert Statistician.

The previous sections dealt with examples in which the person assisted was a statistical novice. Another line of research continues the venerable tradition of software built by statisticians for statisticians. For who else knows better what would help a statistician?

A leading example of this line of research is the system DINDE (French for 'turkey', not an acronym), described by Oldford and Peters (1986b). DINDE aims to ease a statistician's data analysis task by providing a visual map of an analysis. To do so, it builds a structure representing the analysis, adding a bit to the structure with each command.

The commands available in the system are defined as producing one step of the analysis. This may be a plot, the fitting of a model, or the calculation of a statistic. Each such step of the analysis is represented by a node. The node will have a number of attributes which are based on the command used in creating it. For instance, it has a set of suggested commands available. The user is free to give the system any command at any time, but the menu of commands available at a node shows the commands most frequently used when attention is focused on the result of the given command.

When they are created, nodes are shown visually as small rectangles about two centimeters square. In the rectangle is a small graph showing some result of the step. For a graphical command, such as a histogram, the resulting graph itself is shown in low resolution. For a regression fit, a plot of the residuals is shown. The rectangle also has the name of the command and the name of the step if it has been given one. By clicking a mouse on the rectangle the user can get several useful options. One option, *zooming*, expands the rectangle enough to show the full results of the command. Another provides the menu of common next steps.

Three sets of links are maintained between the nodes. These sets are called the *analysis*, the *data flow*, and the *causal* links. Each of these set of links structures the nodes into a set of directed hierarchical graphs. The causal links are created by the computer whenever a new node is created by selection of a command from the common operations menu of an existing node. These links do not change during the analysis. The analysis links are entirely under the users' control, although the computer inserts a default analysis link whenever it creates a causal link. The data flow links show the source of each input to the command represented by the node.

The default graphical presentation of the analysis shows the nodes and the analysis links. On command, one of the other sets of links can be used to structure the display. The user manipulates analysis links to produce a personally meaningful display of the analysis. The default links are usually desired, and analysis links are added to show relations between causal graphs initiated by the user.

As the analysis grows, it becomes impossible to show all the nodes at once. Two similar grouping mechanisms are provided. The user can group any set of nodes into an *analysis map*, which is then shown as a single rectangle on the screen. The links shown to the analysis map are the union of all links from any of the enclosed nodes. When an analysis map is zoomed in on, operations on the nodes can be made, with the resulting nodes incorporated into the map. Any number of levels of analysis maps can be used, so long as the structure remains hierarchical (that is without self referential loops). *Views* are a similar device, but do not allow operations to be made.

DINDE was developed using the Loops(Tm software environment. This is an extension of Lisp that provides support for object oriented programming and for rules. DINDE has relied heavily on object oriented programming, with its view of data analysis clearly influenced by the tool used. Object oriented programming means writing programs to deal with hierarchically structured objects. The operations available for any given object are those defined explicitly for it plus all that it inherits from its superordinate objects. It has proved useful when the concepts that a program needs to deal with have a natural hierarchical structure.

The formalization for description of data used by DINDE, described by Oldford and Peters (1986c), is also quite interesting. An observation is regarded as the *value* of a *variate* for an *individual*. Thus, values, variates, and individuals are the primitives used. Individuals are simple entities, featuring only a name, a description, and comments. Variates contain the same information; in addition, they are categorized as continuous, discrete, or categorical variates. Each variate specifies a range, which can be determined or set externally. The result of a measurement of a variate on an individual is stored in a *datum* object. The object records the number or string obtained from the measurement, the censoring (right, left, or none), and the number of significant digits, if applicable.

For example, suppose a person, George, is known to be taller than 1.71 meters. Then an object is made to represent George if he is not known already, an object is made to represent human height if the variate is not known already, and a datum is created with value 1.71, right censoring, and 3 significant figures. The units for recording human height will be recorded in the variate, along with methods to convert from other units. Notice that if more than one measurement is used for George, it will be easy to determine that the data are paired, and thus to limit suggestions for the kind of analyses to perform.

These primitives are then associated to form higher level constructs. A datum can be associated with a variate, and a group of such associations in turn associated with an individual as a *case*. Or a datum can be associated with an individual, and a group of such associations in turn associated with a variate as a *factor* or a *batch* depending on whether the variate is categorical or not. Many of the operations on cases, factors, and batches are the same, such as locating a datum value for a given identifier, or finding how many datum values are censored, or translating the values into the form of a vector for numerical convenience. By carefully constructing superordinate concepts, such common operations are inherited and only need to be programmed once. Other operations are specific to the kind of association, such as producing boxplots or histograms for a batch, but not for a factor or a case.

This representation is, of course, not directly observable by the user. The reason it is important is that it enables new services to be provided to the user.

2.5. Consultation System Constructors.

Another type of system designed for use by statisticians is one that allows a statistician to build a consultation system (that a novice will use). This sort of system differs from most of the previous ones considered, which intended to help the user directly to perform a data analysis task. A consultation system constructor is designed to help users indirectly by letting a statistician provide them with a consultation system. THESEUS, by providing a rule editor, is a primitive consultation system constructor, or shell for statistical consultation systems. However, it is possible to incorporate more advanced AI concepts into the construction system. The first exploration of this possibility for a consultation system constructor was Student (Gale, 1987), although it was only a feasibility demonstration.

Student was designed to allow a professional statistician to build a knowledge-based consultation system in a data analysis technique by selecting and working examples and by answering questions. The statistician was not supposed to need to know the internal representation of the strategy demonstrated (such as specific rules), and was not supposed to need to know how to write a knowledge based program. He was supposed to be fluent in the underlying statistical system, a more natural expectation of a statistician.

Like REX and other data analysis consultation systems, Student is based on an underlying statistical analysis system, and constitutes an interface to that system. Student used Quantitative Programming

Environment (QPE) (Becker and Chambers, 1986) as the underlying system. Briefly, QPE was designed as the basis for a new version of S (Becker and Chambers, 1984) and has since been incorporated into the "New S" (Becker, Chambers, and Wilks, 1988). The old version of S provided an interactive environment for statisticians, providing data management services, highly portable interactive graphics, command interpretation, and number crunching. In the new S, the external syntax and appearance have been largely maintained, while QPE has added tools such as programming, browsing, debugging, and editing capabilities. The design of Student assumed that the statistician using Student to create a consultation system knew how to use the New S.

A methodological study of Student (Gale 1986c) was built using Lisp and a Symbolics machine. A second version was built in the language provided by S. The goals of the S version were to study issues such as speed, usefulness to statisticians, and generality of the conceptual framework used by Student. This version showed the knowledge acquisition methods more clearly than the Lisp version, and was used to acquire a few different data analysis strategies.

What Student added to the capabilities of REX was the capability to acquire its knowledge base by interview and demonstration. The demonstration approach was proposed by Gale and Pregibon (1984), and tested in the Lisp prototype (Gale 1986c).

The importance of acquiring a strategy by interview and demonstration is considerable. In the current state of building knowledge-based consultation systems, two distinct roles, usually played by two different people, are standard. One is the role of subject matter expert, and the other is the knowledge engineer. In building REX, I played the knowledge engineer, while Daryl Pregibon played the statistical expert. This procedure requires the knowledge engineer to learn a lot about the subject matter, or the subject matter expert to learn a lot about the inference engine and programming, or both.

The primary goal of this kind of software is to allow a statistician, who does not know how the inference engine is built, to build a knowledge based consultation system without the involvement of a knowledge engineer. This should support greater efficiency in building consultation systems in data analysis.

There is a substantial secondary benefit as well. A statistical consultation system will be used in many other ground domains, such as physics, psychology, or business analysis. Current AI techniques are not adequate to handle knowledge in multiple domains, so we built REX with the explicit assumption that the user was willing to learn statistics concepts and vocabulary. This assumption will be reasonable for many analysts, but it will be unreasonable for many managers or low frequency users of statistics.

Another significant benefit of removing dependence on a knowledge engineer is the capability to specialize a system to a local environment. When the construction system is first acquired by a group such as a quality engineering group, a specialist statistician can select examples from the group's files and work them in the system's environment. After this specialization training, the engineering experts would use the system for consultation, returning to the statistician with problems beyond its training. When such a problem seemed frequent, the statistician would work it as an addition to the strategy. If it seemed infrequent, then it would be worked by hand.

The leading example of a consultation shell is currently TAXSY (Darius, 1990), which has reached the stage of a prototype that I was able to demonstrate at the Prague Summer School. Like Student, TAXSY is embedded in an existing statistical language, in its case, SAS. SAS was not designed for symbolic programming, and the production of TAXSY has been a *tour de force* by Darius as an expert in SAS. Even so, use of SAS has made TAXSY clumsy to use, and Darius (private communication) has recently reprogrammed TAXSY using S.

TAXSY is a set of SAS programs designed to work with a set of user supplied structures, each in the familiar form of a SAS data set. The user supplied structures are grouped as strategy, rules, and prompts. Some auxiliary bookkeeping structures are computed by TAXSY to speed access into the primary data sets. As such, the user is clearly aware of the representation of the strategy. This restriction has not yet been overcome by a prototype.

The central part of TAXSY is an inference engine which performs backward chaining on the rules. What is novel about TAXSY is that, by being built within SAS, it has full access to all of the powerful calculations that SAS can make.

TAXSY also allows the end user to take over control of the process, a feature that is to be found in most of the consultation programs now. The end user can also vary the amount of tracing that is done. A session can be interrupted and saved, and then continued at some later time.

3. NON-EXPERT SYSTEM APPLICATIONS OF AI

While the first applications of AI in statistics were clearly from the expert system developments in AI, other areas of AI have begun to be applied, and many more remain to be explored.

3.1. Two Systems.

A system designed *for* experts rather than *as an* expert is KENS, described by Hand (1987). KENS is not a data analysis system at all, but is termed a "Knowledge ENhancement System." It provides paragraphs of text and extensive cross referencing aids so that statisticians can remind themselves about statistical points, or learn about the particular statistical system. Hand suggests that it would be useful to (1) find out *how* to use a system to do something specific one already knows about, such as a Wilcoxon test, (2) be reminded of the available options for a general task one knows one wants to do, such as to compare two groups, and (3) draw one's attention to a missing and crucial piece of information, as in trying to determine if a chi-squared value of 6.2 were significant.

The basic psychological observations behind KENS is that it is easier to recognize than to remember, and that without constant use skill and knowledge are forgotten. The intent of KENS is that an expert relearn some fact rather than that a novice learn the fact. KENS is a non-linear structuring of the same kind of knowledge as one would find structured linearly in a text book.

KENS is built as two sets of nodes. One set contains the paragraphs, another set contains indexing terms. Along with each paragraph there is a brief, summarizing title, and a summarizing list of words and phrases. Links from indexing terms to paragraphs may be labeled definitions or examples, or may be unlabeled. KENS has notions of subconcepts and superconcepts, equivalent concepts, and opposites.

The formalization of knowledge in KENS is rudimentary. The labor of making a system useful in the short term is thus shifted to the statistician who would provide the node contents. If systems such as this are built and proven useful, however, they provide a very interesting beginning to an extensive formalization of statistical knowledge. Such an extensive formalization would seek to refine each paragraph node of a KENS-like system into a machine usable set of propositions. The system might then be able to use natural language generation capabilities to formulate tailored responses to more specific questions.

KENS is a successful prototype. It has been tested by approximately a hundred users, and I demonstrated it at the Prague Summer School.

The other example of an AI application that is not an expert system is EXPLORA (Hoschka and Klosgen, 1990). EXPLORA was designed to extract natural language statements about a data base that are both "interesting" and true. It is written as a shell that can be specialized to different data bases by interested users. It has several key components, some of which must be provided by the user. The first is a set of templates which define the statement types of interest, and associated methods for determining if a particular example of the statement type is true of the data base. These components must be provided by the user, as is the data. EXPLORA then generates a set of true statements from the data base. The paper does not say how this initial set is limited, which must be a problem for large data bases.

Given the set of true statements, EXPLORA then selects a non-redundant subset of the statements. To do so it uses abstractions representing the relations between objects in the data base. The relations are partially provided by the user, and partially by the builders. For instance, the system is being provided for German users and relations of German geography are frequently of interest, so the builders have described these relationships. The relationships can be hierarchical or causal. The selection of non-redundant statements is done by a few simple rules, such as retaining only the statement for the highest node in a hierarchy for which the statement is true.

EXPLORA then uses techniques taken from the natural language generation research in AI to generate a report. The techniques focus on discourse and and sequencing. In some cases an accompanying graphical presentation can also be made.

3.2. Other Possible Applications of AI research.

A major area of AI activity may be called problem solving. It includes planning and consultation systems. The work in consultation systems is the main AI application which does have an analogue in statistics, as we have seen. AI systems which plan programs (Sussman, 1975), molecular genetics experiments (Stefik, 1980), and medical experimental protocols (Weiner, Horwitz and Bauer, 1987) have been built.

One natural analogue in statistics is experimental design. This is an area well studied by statisticians as a formalized discipline of designing matrices with suitable orthogonality properties. The matrices are based on how many factors and how many interactions of factors the experimenter wants to be able to distinguish by analysis of the results of the experiment.

The system described by Haaland and others (1986, 1990) uses AI techniques, but does not move beyond the performance of non-AI systems. In particular, it does not move beyond the formalized view of experiment design as matrix selection. Perhaps more significant aid would require specialization to an industry or company, so that knowledge about the subject of the experiment could be incorporated into the design.

The language for describing experiments developed by O'Keefe is a possible beginning of a more comprehensive view of experiments than matrices of factors. A system might be able to build a good description of an experiment design in interaction with a user. Such a description would be useful both for executing the experiment and for analyzing the results. It should be subject to modification to reflect the accidents that happen during an experiment.

Another area of AI activity has been natural language processing. This means dealing with strings of characters, not sounds. A lot of work has been put in, and the results so far are mostly interesting research which falls into comprehension, generation, and dialogue. Comprehension research deals with paragraph length stories. Generation research has shown ability to use user-oriented vocabularies, and has achieved sentence level utterances. Dialogue research is recent and has just begun to identify the hard problems. EXPLORA shows a kind of statistical application that can benefit from this research.

Currently the most practical application is natural language interfaces for data base queries. The domain of discourse is naturally restricted to the part of the world represented in the data base, which limits the concepts to a manageable number. The requests and replies are each at the sentence level. The systems are easier to use and easier to train people to use than formal data base query languages.

The importance of a report of a statistical analysis suggests a careful study of generation techniques, although at a longer time range, is the following.

Knowledge bases for consultation systems are currently not very readable by people. The current approach is to generate natural language versions of the knowledge from the machine readable version. Another approach would be to devise a restricted natural language that could be compiled into a machine readable knowledge base. The problem is to make the knowledge in an AI system as accessible for comment by experts as is the knowledge in a journal article or book.

There has been a long history of belief that computers would be useful in education. The belief has not been borne out by experience, but there is always the hope that the next generation of computer techniques will turn the trick. Accordingly, there has been research into applying AI ideas to tutoring. One hope has been that a model could be built of the student, which would allow the tutorial material to be specialized for the student. For a system to build a model of the student from observing its interactions with the student, it is necessary for the form of the model to be prespecified. The successful models have been competency models, that is, they show what tasks the student can successfully perform. Models of ways that students can make mistakes have foundered on the multiplicity of ways of going wrong. One of the more successful AI based tutoring systems (Anderson 1984) simply does not allow the student to continue after a mistake, but corrects it immediately.

The possible applications in statistical education are clear. At present, any work done on this would best be done as part of a team which included expertise on educational issues as well as statistical issues. Barzilay (1984) described a system for tutoring in Bayesian concepts of probability.

A final area of AI interest for which I can see an analogue in statistics is formal models of common sense reasoning (Hobbs and Moore, 1985). One motivation for studying these is based on the observation that a person does not solve the Navier-Stokes hydrodynamic equations before jumping out of the way of spilling hot coffee. This behavior suggests that people have rapid and accurate models of liquids. AI research has attempted to build qualitative models using just states (Hayes, 1985), or states and directions of change (de Kleer and Brown, 1985). The models produced so far have only been implemented as research systems; they have not been used in practical systems.

The analogue that I see in statistics is models of common sense understandings of uncertainty and probability. Such models could be based on experimental observations begun by Piaget (1951). Their use would be to allow a system to model its users, and thus to provide explanations understandable at the user's level of development in concepts of uncertainty. This might include some teaching of incrementally advanced concepts.

It is also worth commenting that I see several opportunities to apply AI techniques in the suggestions of Tukey (1983). "Cognostics," for instance, are diagnostics for interpretation by a machine rather than by a human. To select the most interesting scatter plots from a large set would require a cognostic for the interest of a scatter plot. To even begin, one must attempt to formalize the notion of "interesting."

4. COMMENTS ON BUILDING STATISTICAL SYSTEMS

4.1 Comments on tools used in statistical systems.

For those who want to disseminate expertise, implementation of a consultation system is clearly required. However, even those who simply want to develop a formalized strategy to clarify their own thinking need to implement it on a computer. The contribution of AI, as pointed out in Section 1, is new tools for implementing formal systems. Without using these tools, no more progress in strategy should be expected than has been seen in the past, that is, very little. The reason for implementing is to be able to test the strategy. Without testing, the strategy cannot be falsified, and without the possibility of falsification there is no science. Therefore, research using AI concepts requires implementing.

In implementing a strategy for data analysis, it is necessary to connect the AI tools to a statistical package. It appears that this necessity has limited the AI tools that have been found useful in statistical application.

Early work either constructed a new statistical package (for instance, DINDE), or constructed the required KE tools (for instance, REX and Student). My original reason for building the AI tools used in REX was that among the two or three tools I could actually get in 1981, I did not find any that I could use along with S. As more KE tools have become available, this approach may no longer be necessary. However, the key issue is connecting the AI component to an existing statistical package. This requirement is frequently met by low level tools (such as APES, (Wolstenholme and Nelder, 1986)), but is a limitation on higher level tools.

However, as attention turns from research to products, an approach that may be useful is building the required KE tools *in, or as an extension of* the target statistical package. In this setting, the statistical package is probably determined by the organization developing the product. The organization may also have the access to the package required to make extensions to support AI programming additions. Darius (1986) reports using SAS, a prominent statistical system, without modifications to implement simple backchaining rules. WAMASTEX (Dorda and others, 1990) and Student are other examples of this approach.

The tools that have been used successfully at all are still low level tools, such as Loops (used by DINDE) and APES (used by GLIMPSE). These are tools which are firmly embedded in a powerful general purpose language popular for AI style programming (Lisp and Prolog, respectively).

Whatever tools are used, I recommend that they at least all work on one machine. This may seem obvious, but the powerful development environment on a Symbolics tempted me into building the Student prototype on a Symbolics communicating with S on a Vax. Even for research that combination was poor.

4.2 The continuity of work on statistical systems.

I have commented at several points in this paper that work on some system ceased when the author was granted a degree. Furthermore, there is only one case, Marlyn Nys, of a graduate student building on previous work by another person. Her advisor, Paul Darius, has been one of a few people to continue building on thier own work. Bill DuMouchel and Paula Hietala are two others. This is a disappointing pattern, even though progress has been made over the last five years. Possibly a more normal pattern will simply evolve as the subject matures.

However, I think there is a hidden reason for this pattern. I am now convinced that the first requirement for a successful project is a definite user population. The reason I stopped working on these systems is that I had no such definite user population. The techniques we have available allow us to build rather general systems, such as MULREG, or rather specific systems, such as WAMASTEX. However, if the system built does not satisfy fairly closely the needs of its intended users, then it will not be used. One cannot predict the needs of users without knowing them, so knowing the users is the first requirement for a successful system. I do not believe this point has been widely appreciated, so many systems have been started without satisfying it.

5. SPECULATIONS ON THE FUTURE OF AI IN STATISTICS

We now have about ten years of research work applying AI in statistics. Where do we stand? The current status of consulting systems is that numerous feasibility systems have been described in the literature. At least half a dozen have become prototypes, that is, systems used by a few friendly users. There is one major commercial system based on formal models of data analysis, and a few minor ones. Simple techniques are working and have been shown to be effective, as testing of effectiveness has begun. Detailed considerations of parts of an effective consultation system, especially the interface, have appeared; the work is digging into important details. Non-expert system applications of AI have appeared. Considerable progress in the five years since (Gale, 1987) is readily apparent.

I am frequently asked what this work means for the future of statisticians. Any answer is, of course, speculative. Yet the uncertainties people have about this seem to influence their view of the work, so some discussion of the issues is appropriate.

The basic goals of consulting systems are lower cost statistical information and higher productivity in using statistical techniques. These goals are both realistic and unlimited. They are realistic because some steps can be taken in the short run. They are unlimited because there is no clear stopping point; even if a human level of performance is achieved, perhaps more could be done.

Given this basic goal, many statisticians have asked me whether these systems will replace statisticians.

In the next five years, I expect to see several major commercial systems. I also expect tools specifically for building statistical expert systems to become available. A number of applications of AI that are not expert systems should also be expected. The impacts of these systems will be different for users, consulting statisticians, and research statisticians.

Users will not see a big difference. They will perceive the systems available to them as becoming increasingly friendly. The basis of the greater friendliness will not and should not be obvious to them. The most advanced consulting software will handle most of the users statistical problems, all the routine ones.

In a setting where consulting systems are used, then, the consulting statisticians will not see routine problems. They will see unusual problems. Their activities will thus include more learning as they consult the large knowledge base of statistics represented in books and journals. They will also advise the users on the availability and choice of consulting systems. To the extent that the software achieves its goal of providing greater productivity, there must be less involvement by the consulting statistician with each user. This reduction in demand for statistician's services may be offset, or more than offset, by more users of the statistical software.

Research statisticians should simply see an increase in the options for research. In the first place, there will be techniques that would not be feasible without formal models. Just as the availability of cheaper computing has made computation intensive techniques such as bootstrapping available, so the availability of smarter computing will make new techniques available. For the foreseeable future, there will remain less common techniques and less common problems of common techniques to formalize. There will need to be many formalizations of common techniques in order to discuss their advantages and disadvantages. Consulting systems will provide the option to embed some research now carried out. I see only more work, not less, for research statisticians.

If TAXSY or some tool like it becomes easily available, there will be some additional impacts, primarily expanding consulting statisticians' roles. The research statistician would find building consulting system or strategy easier, and would thus be encouraged to do so. The consulting statistician would have an important new role in specializing the systems to their local environment. This would be a continuing activity as well as an initial one, because they would be able to encode the techniques they learned for problems encountered by their users. Users would find the software more specialized to their environment.

I see three limits to any consulting system software in the foreseeable future. First, the systems will have very limited knowledge compared to the entire corpus of statistical knowledge. A good programmer, working with known algorithms and a specified performance, can write programs with 10^6 bits in a year. With diseconomies of scale, unknown methods, and unclear goals, I would not expect that much productivity in writing consulting systems in the next five to ten years. A crude estimate of the extent of the body of statistics knowledge is 10^{11}: 1000 library shelves, each with 30 volumes, each of 10^6 bits, with another small factor for closely allied areas. Thus only a fraction of the knowledge will be encodable.

Computer systems will not have access to the body of knowledge as represented in the books and journals for the foreseeable future. The work on natural language understanding has shown the huge knowledge that is necessary in order to learn a little bit more. Writing, even in journal articles, is remarkably informal, with frequent appeals to common experience for motivation. Learning techniques in AI are still rudimentary despite a great interest in the topic since the earliest days of AI; this is a hard problem.

The time required for a person to digest the results of an autonomously running program will remain small. John Tukey (1986) pointed out the huge number of cycles available on computers every night (ten hours is 4×10^{10} microseconds). It would be desirable to have systems that could use this time autonomously, providing output for the statistician to examine in the morning. As of now, a program that a statistician never tended could interest a statistician for only a few minutes each morning. I think an interesting autonomous program would necessarily have to be a learning program, so I do not see much progress on this in the near future.

Since Artificial Intelligence is now only about 40 years old, I do expect progress over a span of several decades on these harder problems. In the mean time, the tools already available make some exciting new options for research by statisticians. And it will be worth watching AI successes in other domains for possible analogues in statistics.

ACKNOWLEDGEMENT

Author wish to thank to Professor Ivo W. Molenaar for carefull reading the manuscript and valuable remarks.

REFERENCES

TOPICAL INDEX TO REFERENCES

Review and General.

Chowdhury S. I. (1987)
Gale W. A. (ed.) (1986 a)
Gale W. A. (1986 e)
Gale W. A. (1987)
Hahn G. J. (1985)
Hand D. J. (1986 b)
Molenaar I. W. (1988)
Streitberg B. (1988)
Wittkowski K. M. (1990)

Discussions of Consultations and Interfaces.

Clayden A. D. and Croft M. R. (1990)
Hietala P. (1992)
Nelder J. A. (1988)
Ollivier M. R., Arrus M., Durillon S. R. and Debord B. (1992)

Knowledge Representation and Acquisition.

Aikins J. S. (1983)
de Kleer J. and Brown J. S. (1985)
Gale W. A. (1986 d)

Gale W. A. and Lubinsky D. (1986)
Hand D. J. (1986 a)
Hand D. J. (1992)
Herrmann N. A., Silvers K., Godfrey B., Roberts B. and Cerys D. (1991)
Hayes P. (1985)
Hobbs J. and Moore R. (1985)
Newell A. (1981)
O'Keefe R. (1985)
Oldford W. and Peters S. (1986 a)
Oldford W. and Peters S. (1986 c)
Oldford W. (1990)
Piaget J. and Inhelder B. (1951)
Pregibon D. (1986)
van den Berg G. M. and Visser R. A. (1990)

Systems on Planning and Choice.

Anderson J., Farrell R. and Sauers R. (1984)
Athey S. (1989)
Barzilay A. (1984)
Blum R. L. (1982)
Dambroise E. (1987)
Haaland P. D., Yen D. and Liddle R. F. (1986)
Haaland P. D., Lusth J. C., Liddle R. F. and Curry J. W. (1990)
Molenaar I. W. and Passchier P. (1992)
Portier K. M. and Lai P. (1983)
Stefik M. J. (1980)
Sussman G. J. (1975)
Weiner J. M. and Horowitz R. (1987)

Systems on Strategy of Data Analysis.

Bell E. and Watts P. (1988)
Bell E. (undated)
Berzuini C., Ross G. and Larizza C. (1986)
Carlsen F. and Heuch I. (1986)
Darius P. (1986)
Dorda W. K., Froeschl A. and Grossman W. (1990)
Duijens I. J., Delft T. J. and van den Berg G. M. (1988)
DuMouchel W. (1990)
Drapier P. (1987)
Gale W. A. (1986 b)
Hietala P. (1988)
Hietala P. (1990)
Lubinsky D. and D. Pregibon (1987)
Raes J. F. M. (1992)
Ross G. J. S. (1988)
Wolstenholme D. and Nelder J. (1986)
Wittkowski K. M. (1988)

Systems to Aid Statisticians.

Darius P. (1990)
Gale W. A. (1986 c)
Gale W. A. (1987)
Gale W. A. and Pregibon D. (1984)
Oldford W. and Peters S. (1986 b)
Prat A., Edmonds E., Catot J., Lores J., Galmes J. and Fletcher P.(1992)

Other Applications of AI.

Hand D. J. (1987)
Hand D. J. (1990)
Hoschka P. and Klosgen W. (1990)

Statistics.

Becker R. A. and Chambers J. M. (1984)
Becker R. A. and Chambers J. M. (1986)
Becker R. A., Chambers J. M. and Wilks A. R. (1988)
Mallows C. L. (1973)
Miller R. G. (1986)
Tukey J. W. (1983)
Tukey J. W. (1986)
Velleman P. F. and Hoaglin D. C. (1981)

[1] Aikins J. S., *Prototypical Knowledge for Expert Systems*, Artificial Intelligence **120** (1983), 163 – 210. (This paper describes Centaur, Aikins' thesis project. It provided a single framework that combined both rules and frames.)

[2] Anderson J., Farrell R. and Sauers R., *Learning to Program in LISP*, Cognitive Science **8** (1984), 87 – 129. (The paper describes an intelligent tutoring system which taught some basics of programming in LISP.)

[3] Athey S., *SAM: Statistics Analysis Mentor*, PC AI, The Artificial Intelligence Magazine for Personal Computing (November/December 1989), 56 – 60. (SAM was designed to assist a novice in making a choice of technique. It uses a decision tree to guide the user, and makes no use of data. This is old and simple technology, but the article reports a test of the usefulness of the software: comparing two groups of naive subjects, one with SAM, and the other with a textbook, the computer group chose 38/60 correct solutions while the textbook group scored 27/66.)

[4] Barzilay A., *An Expert System for Tutoring Probability Theory*, PhD Thesis submitted to Graduate School of Business, University of Pittsburgh, 1984. (The system SPIRIT was built using OPS5. It tutors in conditional probabilities.)

[5] Becker R. A. and Chambers J. M., *S: an Interactive Environment for Data Analysis and Graphics*, Wadsworth, Belmont, California, 1984. (This is the manual for the use of (old) S, a statistical computing system produced by AT&T Bell Laboratories. It contains information on ordering *S*.)

[6] Becker R. A. and Chambers J. M., *Auditing of Data Analysis*, Proceedings the American Statistical Association, Statistical Computing Section, American Statistical Association, Arlington, Virginia, 1986, pp. 11 – 18. (The paper describes the use of the new *S* in a task of use to statisticians.)

[7] Becker R. A., Chambers J. M. and Wilks A. R., *The New S Language*, Wadsworth, Pacific Grove, California, 1988.

[8] Bell E., *THESEUS*. (A guide to installation and use of the expert system and rule base editor. Development of THESEUS ceased when Bell received her doctorate, but I was able to demonstrate the prototype at the Prague Summer School using her program and documents.)

[9] Bell E. and Watts P., *Building a Statistical Knowledge Base: A Discussion of the Approach used in the Development of THESEUS, a Statistical Expert System*, In: COMPSTAT'88: Proceedings in Computational Statistics (Edwards D. and Raun N., eds.), Physica-Verlag, Heidelberg, 1988, pp. 143-148. (A version of THESEUS was demonstrated at this meeting. Bell found that each statistician she surveyed had a different strategy, and built a system that would allow them to incorporate their own expertise into the rule base.)

[10] van den Berg G. M. and Visser R. A., *Knowledge Modeling for Statistical Consultation Systems: Two Empirical Studies*, In: COMPSTAT'90: Proceedings in Computational Statistics (Momirovič K. and Mildner V., eds.), Physica-Verlag, Heidelberg, 1990, pp. 49-56. (This paper reports the first experimental work to study the methods used by statisticians, and the factors they consider to be important in chosing among the methods. Considerable agreement between the statisticians was found in classifying methods for exploratory or confirmatory analysis, and in classifying methods as suitable for discrete or continuous variables, and in grouping methods based on generalized linear models. However, there was no agreement on the classification of methods not based on generalized linear models. A principal components analysis showed a first dimension of methods with dependent variables versus those without, and a second dimension of methods for one or two variables versus multiple variables.)

[11] Berzuini C., Ross G. and Larizza C., *Developing Intelligent Software for Non-Linear Model Fitting as an Expert System*, In: COMPSTAT'86: Proceedings in Computational Statistics (De Antoni F., Lauro N.C. and Rizzi A., eds.), Physica-Verlag, Heidelberg, 1986. (This is a first paper on a project to make an intelligent interface to the *Maximum Likelihood Program*, a program encoding an advanced statistical procedure.)

[12] Blum R. L., *Discovery and Representation of Causal Relationships from a Large Time-Oriented Clinical Database: The RX Project*, Springer-Verlag, New York, 1982. (The book is based on Blum's thesis. It describes the whole RX system.)

[13] Carlsen F. and Heuch I., *EXPRESS — An Expert System Utilizing Standard Statistical Packages*, In: COMPSTAT'86: Proceedings in Computational Statistics (De Antoni F., Lauro N.C. and Rizzi A., eds.), Physica-Verlag, Heidelberg, 1986, pp. 265-270. (This system compared the locations of two samples. It used BMDP for calculations, and introduced simulation as a means of testing the system.)

[14] Chowdhury S. I., *State of the Art in Statistical Expert Systems*, 7th International Workshop: Expert Systems & Their Applications, Avignon, France, 1987. (This paper has half page summaries of 17 previous papers.)

[15] Clayden A. D. and Croft M. R., *Statistical Consultation — Who's the Expert?*, Annals of Mathematics and Artificial Intelligence **2** (1990), 65-76. (The authors collected data on actual consultation sessions over a three year period. They conclude that *The simplistic view of an expert system using almost trans-*

parent rules of statistics clearly does not fit the characteristics of research consultation.)

[16] Dambroise E., *MUSE: Multivariate Expertise*, PhD Thesis prepared under Y. Escoufier, INRA, 9 Place Pierre Viala, 34060 Montpelier Cedex, France. (The thesis provides Dambroise with enough space to explain MUSE in detail, but the paper is much easier to get.)

[17] Dambroise E. and Massotte P., *MUSE: an Expert System in Statistics*, In: COMPSTAT'86: Proceedings in Computational Statistics (De Antoni F., Lauro N.C. and Rizzi A., eds.), Physica-Verlag, Heidelberg, 1986, pp. 271-276.

[18] Darius P., *Building Expert Systems with the Help of Existing Statistical Software*, In: COMPSTAT'86: Proceedings in Computational Statistics (De Antoni F., Lauro N.C. and Rizzi A., eds.), Physica-Verlag, Heidelberg, 1986, pp. 277-282. (Darius used *SAS*® as the programming language for a rule interpreter, and used that to select a test for comparing two group means.)

[19] Darius P., *A Toolbox for Adding Knowledge-Based Modules to Existing Statistical Software*, Annals of Mathematics and Artificial Intelligence **2** (1990), 109-116. (TAXSY is an outgrowth of the work described in Darius (1986). It is the only available consultation system shell for statistical systems. Statistical expert systems all face the problem of connecting knowledge-based choices to computation. TAXSY takes the approach of building a shell within the *SAS*® statistical system. The work has reached the stage of a prototype, as I was able to demonstrate it at the Prague Summer School. However, the use of *SAS*® for symbolic programming is quite clumsy, and Darius is now transporting TAXSY to *S*.)

[20] Dorda W., Froeschl K. A. and Grossman W., *WAMASTEX — Heuristic Guidance for Statistical Analysis*, In: COMPSTAT'90: Proceedings in Computational Statistics (Momirovič K. and Mildner V., eds.), Physica-Verlag, Heidelberg, 1990, pp. 93-98. (WAMASTEX 1 is a prototype in use within the Vienna University Hospital. The authors had a clear group of users to design for and made a system with the limited goals of univariate description. Heuristics were tested experimentally by getting responses from 37 practicing statisticians. The program is integrated into *SAS*®. Further versions are being designed.)

[21] Drapier P., *Le systeme de Regression Assistee par Ordinateur*, preprint, RAO, 20 Rue Rouget de L'Isle, 94100 Saint Maur, France, 1987. (The system described runs on a microcomputer and tackles multiple linear regression.)

[22] Duijens I. J., Delft T. J. and van den Berg G. M., *PRINCE: An Expert System for Nonlinear Principal Components Analysis*, In: COMPSTAT'88: Proceedings in Computational Statistics (Edwards D. and Raun N., eds.), Physica-Verlag, Heidelberg, 1988, pp. 149-153. (PRINCE is an example of a program designed with expert system principles to prepare input statements for an older program with arcane input requirements. It was demonstrated at the conference by the developers. A copy sent to me while I was preparing for the Prague Summer School performed as its documentation described, but I did not demonstrate it at the school as its output was not readily comprehensible.)

[23] DuMouchel W., *The Structure, Design Principles, and Strategies of MULREG*, Annals of Mathematics and Artificial Intelligence **2** (1990), 117-134. (MULREG is the most fully developed statistical expert system. The paper describes how graphics were integrated into each stage, and the choice of statistical algorithms to minimize interactions and thus allow as general a strategy

as possible.)

[24] Gale W. A. (ed.), *Artificial Intelligence and Statistics*, Addison Wesley, Reading, Massachusetts, 1986 a. (This is the first book published in the field. It is a logical starting point for learning about the field.)

[25] Gale W. A., *REX Review*, pp. 173 – 228 in [2], 1986 b. (This was the last and longest of a series of papers on REX. The papers before this do not add much. It describes what REX looked like when it ran, gives an example of the report generated by REX, and describes the inference engine and knowledge representation in detail. It discusses the statistical strategy only briefly.)

[26] Gale W. A., *Student Phase 1: A Report on Work in Progress*, pp. 239 – 266 in [2], 1986 c. (This described the first version of Student, implemented on a Symbolics Lisp Machine.)

[27] Gale W. A., *Knowledge – Based Knowledge Acquisition for a Statistical Consulting System*, International Journal of Man – Machine Studies **26** (1986 d), 55 – 64. (This paper describes the abstract Knowledge Engineering principles on which Student is based.)

[28] Gale W. A., *Overview*, Chapter 1 in [2], 1986 e. (The chapter reviews all papers that I knew about at the time, because they were only available in conference proceedings.)

[29] Gale W. A., *Statistical Applications of Artificial Intelligence and Knowledge Engineering*, The Knowledge Engineering Review **2** (1987 a), 227 – 247.

[30] Gale W. A., *STUDENT: A Tool for Constructing Consultation Systems in Data Analysis*, Proceedings of the 46th Session of the International Statistics Institute, ISI, Tokyo, 1987 b. (This described the second version of Student, implemented in the new S statistical language.)

[31] Gale W. A. and Lubinsky D., *A Comparison of Representations for Statistical Strategies*, Proceedings the American Statistical Association, Statistical Computing Section, American Statistical Association, Arlington, Virginia, 1986, pp. 88 – 96. (The paper compares the knowledge representations used by Student and TESS.)

[32] Gale W. A. and Pregibon D., *Constructing an Expert System for Data Analysis by Working Examples*, In: COMPSTAT'84: Proceedings in Computational Statistics (Havránek T., Šidák Z. and Novák M., eds.), Physica – Verlag, Vienna, 1984, pp. 227 – 236. (This paper described the motivation and plans for building STUDENT.)

[33] Haaland P. D., Yen D. and Liddle R. F., *An Expert System for Experimental Design*, Proceedings the American Statistical Association, Statistical Computing Section, American Statistical Association, Arlington, Virginia, 1986, pp. 78 – 87. (The paper describes a system which helps an industrial engineer select an experimental design.)

[34] Haaland P. D., Lusth J. C., Liddle R. F. and Curry J. W., *DEXTER: A Guide to Selecting the Best Design for an Industrial Screening Experiment*, Annals of Mathematics and Artificial Intelligence **2** (1990), 179 – 196. (The prototype described in the previous paper was developed into a product used internally in Becton Dickson. In doing so, many of the AI concepts originally tried were found superfluous.)

[35] Hahn G. J., *More Intelligent Statistical Software and Statistical Expert Systems: Future Directions*, The American Statistician **39** (1985), 1 – 16. (This was the first major review. It is directed to statisticians, and contains sugges-

tions for research directions.)

[36] Hand D. J., *Patterns in Statistical Strategy*, pp. 355 – 388 in [2], 1986 a. (This paper gives a theoretical view of the task undertaken by the various systems referenced below under strategy of data analysis.)

[37] Hand D. J., *Expert Systems in Statistics*, The Knowledge Engineering Review **1** (1986 b), 2 – 10. (Hand's papers typically have a considerable theoretical framework to which their substance is related. This review paper is no exception, doing a nice job of describing aspects of the statistical domain that the KE should appreciate.)

[38] Hand D. J., *A Statistical Knowledge Enhancement System*, preprint, Institute of Psychiatry, London. (This was the first paper describing KENS, a system to act as an on line reference book.)

[39] Hand D. J., *Practical Experience in Developing Statistical Knowledge Enhancement Systems*, Annals of Mathematics and Artificial Intelligence **2** (1990), 197 – 208. (KENS was tested in dozens of places (and I demonstrated it in the Prague Summer School). The comments of the users on this prototype led to initial work on two other systems.)

[40] Hand D. J. (ed.), *Measurement Scales as Metadata*, In: Artificial Intelligence Frontiers in Statistics, 1992. (Metadata are information about the variables which occur during a statistical study. One particular type of metadata is scale of measurement, such as categorical or continuous. The paper considers this type of metadata in some detail and shows how it can be used in statistical expert systems.)

[41] Hayes P., *Naive Physics I: Ontology for Liquids*, pp. 71 – 108 in [22], 1985. (The paper introduces a formalization for the behavior of liquids.)

[42] Herrmann N., Silvers A., Godfrey K., Roberts B. and Cerys D., *Techniques and Issues in the Development of a Statistical Strategy for a Statistical Advisory System*, preprint (1991), Drexel University Department of Mathematics and Computer Science. (The paper describes how a strategy was developed for a front end for the PROPHET system, a national biostatistical resource used by about 1000 researchers at 40 institutions. The core of the approach was to use the strategy published by R. G. Miller Jr. in *Beyond ANOVA*. The core was augmented as needed for the specific group of users.)

[43] Hietala P., *Inside a Statistical Expert System: Statistical Methods Employed in the ESTES System*, In: COMPSTAT'88: Proceedings in Computational Statistics (Edwards D. and Raun N., eds.), Physica – Verlag, Heidelberg, 1988, pp. 163 – 168. (This paper was the first to describe ESTES, an expert system for time series analysis. It was a feasibility demonstration at this point. It was intended to provide guidance for inexperienced users. The statistical strategy for a particular property considered the granularity of the analysis process desired, the goal of the analysis, and the user's experience.)

[44] Hietala P., *Enhancing Explanation Capabilities of Statistical Expert*, In: Artificial Intelligence Frontiers in Statistics Systems through Hypertext (Hand D. J., ed.), 1992. (The most successful statistical expert systems builders have typically found that explanation was an essential feature. This paper discusses how hypertext, which can combine figures and text, can enhance explanations.)

[45] Hietala P., *ESTES: A Statistical Expert System for Time Series Analysis*, Annals of Mathematics and Artificial Intelligence **2** (1990), 221 – 235. (This paper shows a sample session of how the prototype stage of ESTES worked. The pro-

gram was in use by some students at the University of Tampere. The program runs on a Macintosh, or I would have demonstrated it at the Prague Summer School.)

[46] Hobbs J. and Moore R., *Formal Theories of the Commonsense World*, Ablex, Norwood, New Jersey, 1985. (This book is a good introduction to building formal theories in areas not previously formalized.)

[47] Hoschka P. and Klosgen W., *EXPLORA: Content Interpretation of Data*, preprint, Instituter for Applied Information Technology, German National Research Center for Computer Science, D-2505 Sankt Augustin. (EXPLORA is a shell for the interpretation of data. That is, users must first provide domain knowledge about the objects of interest and the kinds of facts which are desired to be reported. Explora uses the fact templates to generate the set of true facts, then applies rules based on the structure, such as a hierarchy, of the data to sift out "interesting" facts. Beta test versions were in use as of the writing.)

[48] de Kleer J. and Brown J.S., *A Qualitative Physics Based on Confluences*, Hobbs and Moore, 1985, pp. 109-194. (This paper introduces a calculus of directions, and applies it in a few examples.)

[49] Lubinsky D. and Pregibon D., *Data Analysis as Search*, submitted to Econometrica (1987). (TESS, the system described, aims to model the *results* obtained by statisticians in analyzing data, not the methods they use. The example studied is regression.)

[50] Mallows C.L., *Some comments on C_p*, Techmometrics **15** (1973), 661-667. (This paper defined C_p.)

[51] Miller R.G., *Beyond Anova*, Wiley, New York, 1986.

[52] Molenaar I.W., *Statistical Consultants and Statistical Expert Systems*, In: COMPSTAT'88: Proceedings in Computational Statistics (Edwards D. and Raun N., eds.), Physica-Verlag, Heidelberg, 1988, pp. 187-192. (Molenaar considers the kinds of interactions that occur in a statistical consultation as a background for discussing roles that a computerized system might play. He points out that a person should not expect to get as much help from a program as they can from a human expert. The technical knowledge of the statistician *might* be incorporated into a knowledge base, but the subtle dialogue of client and statistician cannot be incorporated now.)

[53] Molenaar I.W. and Passchier P., *The AGREE consultant: More Agreeable Software?*, Journal of Applied Statistics (1992). (The paper describes a front end designed to assist a novice in choosing between the nearly three dozen possible coefficients expressing agreement among experts in a classification task which are produced by a preexisting program, AGREE. The unique feature of the work is that four versions of the program were tested on a sample of users who knew neither the AGREE program nor the theory behind the various agreement coefficients. Correct recommendations were obtained in 395 of 416 trials.)

[54] Nelder J.A., *How Should the Statistical Expert System and its User See Each Other?*, In: COMPSTAT'88: Proceedings in Computational Statistics (Edwards D. and Raun N., eds.), Physica-Verlag, Heidelberg, 1988, pp. 107-116. (This paper describes the interaction capabilities of GLIMPSE (see Wolstenholme and Nelder, 1986, below). The discussion moves somewhat beyond the specific system, and emphasizes that the interface is at least as important as the expertise.)

[55] Newell A., *The Knowledge Level*, AI Magazine **2** (1981), 1-20. (His presidential

address introduced an idea that has organized further research.)

[56] O'Keefe R., *Logic and Lattices for A Statistical Advisor*, University of Edinborough., University of Edinborough., 1985. (PhD Thesis prepared under Alan Bundy at University of Edinborough. This thesis describes a system that selects statistical analysis techniques, but the greater value is a beginning on a formalization of data description and experiment design.)

[57] Oldford W., *Software Abstraction of Elements of Statistical Strategy*, Annals of Mathematics and Artificial Intelligence **2** (1990), 291 – 307. (Just as floating point numbers in a computer are abstractions approximating real numbers, we need software abstractions approximating statistical concepts. The paper discusses the representation of the concepts apart from what action may be taken on them.)

[58] Oldford W. and Peters S., *Implementation and Study of Statistical Strategy*, pp. 335 – 353 in [2], 1986 a. (This paper gave a theoretical view of strategy for data analysis that is still useful to read.)

[59] Oldford W. and Peters S., *Data Analysis Networks in DINDE*, Proceedings the American Statistical Association, Statistical Computing Section, American Statistical Association, Arlington, Virginia, 1986 b, pp. 11 – 18. (This paper describes DINDE's behavior and the knowledge representation behind it.)

[60] Oldford W. and Peters S., *Object Oriented Data Representations for Statistical Data Analysis*, In: COMPSTAT'86: Proceedings in Computational Statistics (De Antoni F., Lauro N.C. and Rizzi A., eds.), Physica – Verlag, Heidelberg, 1986 c, pp. 301 – 308. (This paper describes the knowledge representation ideas behind DINDE.)

[61] Ollivier M., Arrus R., Durillon M., Robert S. and Debord B., *AMIA – Expert System for Simulation Modeling and Sectoral Forecasting*, In: Artificial Intelligence Frontiers in Statistics (Hand D.J., ed.), 1992. (This is a preliminary report from one of the long range DOSES projects. Well established models are used for long range energy demand forecasts, however the models are difficult to use. The project will focus on providing a front end to these models.)

[62] Piaget J. and Inhelder B., *La gene'se de L'ide'e de hasard chez l'enfant*, Presses Universitaires de France, 1951 ((French)); English translation by Leake, Burrell, and Fishbein, *The Origin of the Idea of Chance in Children*, Norton, New York, 1975. (A book that describes experiments showing the developmental stages of ideas of randomness and order.)

[63] Portier K. M. and Lai P., *A Statistical Expert System for Analysis Determination*, Proceedings the American Statistical Association, Statistical Computing Section, American Statistical Association, Arlington, Virginia, 1983, pp. 309 – 311. (This paper describes the behavior of a system based on a decision tree.)

[64] Prat A., Edmonds E., Catot J., Lores J., Galmes J. and Fletcher P., *An Architecture for Knowledge Based Statistical Support Systems*, In: Artificial Intelligence Frontiers in Statistics (Hand J., ed.), 1992. (This paper reports on a long range project to develop tools for the construction of knowledge – based front – ends. The tools had been used to develop two front ends as of the writing.)

[65] Pregibon D., *A DIY Guide to Statistical Strategy*, pp. 389 – 400 in [2], 1986. (This is a collection of rules of thumb on knowledge acquisition from the first statistician to engage in formalizing a strategy.)

[66] Raes J. F. M., *Inside Two Commercially Available Expert Systems*, In: Artificial

Intelligence Frontiers in Statistics (Hand D. J., ed.), 1992. (Raes did a reverse engineering job on SPRINGEX, and STATISTICAL NAVIGATOR, two systems offered commercially and described as expert systems. He presents the complete strategy that each uses, showing their simplicity. The duplication of these strategies would make useful exercises for a course in statistical expert systems.)

[67] Ross G. J. S., *Expert Systems for Non-Linear Modeling: Progress and Prospects*, In: COMPSTAT'88: Proceedings in Computational Statistics (Edwards D. and Raun N., eds.), Physica-Verlag, Heidelberg, 1988, pp. 155-161. (This paper described a feasibility demonstration for an expert system in non-linear modeling which was demonstrated at the conference.)

[68] Stefik M. J., *Planning with Constraints*, Report No. 80-784 (Doctoral thesis), Computer Science Department, Stanford University, 1980. (This thesis shows how knowledge about molecular genetics can be used to constrain plans and reduce an intractable search space.)

[69] Streitberg B., *On the Nonexistence of Expert Systems - Critical Remarks on Artificial Intelligence in Statistics*, Statistical Software Newsletter **19** (1988), 55-62. (The article is followed by comments by J. M. Chambers, W. A. Gale, D. Pregibon, P. Hájek, T. Havránek and J. Nelder, pp. 63-72, and a rejoinder by Streitberg, pp. 73-74. The core of the article is a proof that some basic classes of rule based systems are undecideable, and therefore no algorithm can guarantee a result. The introduction to this unarguable result made several more controversial claims, which were challenged by the respondents. The article forcefully raised the issue of how much progress was being made toward commercially available systems.)

[70] Sussman G. J., *A Computer Model of Skill Acquisition*, Elsevier, New York, 1975. (This book described a program that would learn to write better programs by analyzing bugs in its programs.)

[71] Tukey J. W., *Another Look at the Future*, In: Computer Science and Statistics, Proceedings of the Fourteenth Symposium on the Interface (Heiner, Sachs, and Wilkinson, eds.), Springer-Verlag, New York, 1983, pp. 1-8. (This paper poses a dozen challenging tasks for statistical computing, some of which may be feasible with KE methods.)

[72] Tukey J. W., *An Alphabet for Statisticians' Expert Systems*, pp. 401-409 in [2], 1986. (This paper appraises the strengths and weaknesses of the idea of applying KE techniques in statistics.)

[73] Velleman P. F. and Hoaglin D. C., *Applications, Basics and Computing for Exploratory Data Analysis*, Duxbury Press, North Scituate, Massachusetts, 1981. (This book provides practical computer programs and the basic ideas of exploratory data analysis.)

[74] Weiner J. M. and Horowitz R., *An Expert System for Analysis of Clinical Trial Data*, In: Computer Science and Statistics, Proceedings of the Nineteenth Symposium on the Interface (Heiberger R., ed.), North-Holland, Amsterdam, 1987. (This paper is not so much on *an* expert system as on the person/machine organization of the large body of knowledge required to design ethical, legal, and knowledge enhancing experiments on treatment of diseases.)

[75] Wittkowski K. M., *Building a Statistical Expert System with Knowledge Bases of Different Levels of Abstraction*, In: COMPSTAT'88: Proceedings in Computational Statistics (Edwards D. and Raun N., eds.), Physica-Verlag, Heidel-

berg, 1988, pp. 129–134. (Wittkowski presents a theoretical framework giving six levels for a statistical expert system. The levels are interaction, domain, strategy, semantics, access, and execution. Since natural language interaction is the subject of active research by a large community, it is not ready for application in systems. Very few systems have attempted to incorporate domain knowledge. Relying on a commercially available system for execution, Wittkowski has worked on a system called PANOS–ES that focuses on strategy, semantics, and access (how to send calls to the execution level). He distinguishes his work from others working on these same levels in that "decisions are primarily based on hypothetical, theoretical, and observable relations," as opposed to empirical distributions in the data.)

[76] Wittkowski K. M., *Statistical Knowledge–Based Systems — Critical Remarks and Requirements for Approval*, In: COMPSTAT'90: Proceedings in Computational Statistics (Momirovič K. and Mildner V., eds.), Physica–Verlag, Heidelberg, 1990, pp. 49–56. (This paper makes a definition of a statistical knowledge–based system, and considers the issues of availability, possibility, and testing raised by Streitberg in [8].)

[77] Wolstenholme D. and Nelder J., *A Front End for GLIM*, In: Expert Systems in Statistics (Haux R., ed.), Gustav Fischer, Stuttgart and New York, 1986. (The paper describes early work on GLIMPSE, a front end for the popular statistical system GLIM. The book cited has about ten other papers, but they were written at a planning stage rather than at a stage with systems developed.)

AT&T Bell Laboratories, 600 Mountain Avenue, PO Box 636, Murray Hill, NJ, 07974-0636, USA; tel: (+908) 582 21 49, fax: (+908) 582 33 40.

E-mail: GALE@RESEARCH.ATT.COM

Batch number: 10725973

Printed by Printforce, the Netherlands